The Neurobiology of Orthodontics

Margaritis Z. Pimenidis

The Neurobiology of Orthodontics

Treatment of Malocclusion Through Neuroplasticity

Springer

Dr. Margaritis Z. Pimenidis
Marathonos Street 22
152 33 Halandri, Athens
Greece
marpimenidis@yahoo.gr

ISBN: 978-3-642-00395-0 e-ISBN: 978-3-642-00396-7

DOI: 10.1007/978-3-642-00396-7

Springer Dordrecht Heidelberg London New York

Library of Congress Control Number: 2009926011

Cover design: eStudio Calamar S.L.

Printed on acid-free paper

Springer is part of Springer Science+Business Media (www.springer.com)

I dedicate this volume to my son Alexandros and his wife Sophia for their unfailing love, help, and support.

I am grateful to Dr. Anthony A. Gianelly and to the late Dr. Melvin L. Moss of Boston University and Columbia University, respectively, two great teachers who have supported my orthodontic and oral biology education and who kindled my interest in this field.

My sincere thanks go to my friends Dr. Eugene Eagles and Dr. Paul J. Batastini, as well as to Dr. Robert L. Vanarsdall who kindly reviewed the drafts.

Memory itself was identified as a discrete mental faculty in ancient Greece. It could be cultivated and trained like other human skills to generate knowledge through learning. Aristotle (384–322 BC) summarized these mental activities thus:

> There is an innate faculty of discrimination in all men whereby we perceive sense objects. In some the sense-perception persists, while in others it does not. Where it does not, there is either no cognition at all outside the act of perception, or no cognition of those objects of which perception does not persist. Where sense-perception does persist, after the act of perception is over, the percipients can still retain the perception in the soul (mind). If this happens repeatedly, a distinction immediately arises between those men which derive a coherent impression from the persistence of sensation and those in which it does not persist. Thus sense-perception (aesthesis) gives rise to memory (mneme), as we hold; and repeated memories of the same thing give rise to experience (empeiria), because the memories, though numerically many, constitute a single experience. Accordingly, those that acquire the faculty of memory are more intelligent and are capable of learning than those which cannot remember. Experience in turn, is the starting point of art and science; art in the world of making and science in the world of facts. Thus, the faculty of art and science are neither innate as determinate and fully developed, nor derived from other developed faculties on a higher plane of knowledge; they arise from sense-perception and cognition [53].

Such in brief outline was the theory of Aristotle of the relationship between perception and cognition; and the theory in its essentials holds good today, as described in Chap. 4.

In the world of orthodontics, this book, focusing on the neurobiology of the orofacial structures and the interaction of the orofacial sensory input with the brain, is a unique and substantive departure from the usual "nuts and bolts" of contemporary orthodontics. It may both educate and excite the practitioner. The education is in the form of a synopsis of the most current information of brain function and research, highlighting the ability of sensory input to modulate and change brain activity. Special emphasis is placed on the orofacial sensory network. The excitement comes from opening new and creative pathways to examine the effects of orofacial sensory input in the development of malocclusions of the teeth and the effects of orthodontic treatment on the brain. This book is recommended reading for all orthodontists.

Boston, MA Anthony Gianelly

Preface

According to the traditional orthodontic view the sensory and motor functions of the mouth are considered as phenomena coded in the DNA molecule of the neuron, and this coding acts as a template, a blueprint or a program, containing all the necessary information for normal oral behavior development, with the corollary that evident oral functional disorders, such as seen in certain malocclusions of the teeth are interpreted as DNA coding errors, as mistakes in some inborn or congenital programs [1, 206].

Contrary to this traditional teaching, however, we now know that sensory experiences can modify the structure and function of the brain or structures within it. In other words, the brain is structured by the senses, through patterned sensory input, along with the many different kinds of chemical molecules produced in the brain, such as neurotransmitters, growth factors, and trophic factors. These chemicals mediate the transmission of information and the changes in the brain, control the growth and survival of neurons, as well as the formation of appropriate neural circuits for the communication between the neurons and processing of information in the cerebral cortex. The changes in the brain, in turn, underlie sensation–perception, motor behavior, mentation, and memory of new learned experiences.

The oral sensory experiences can influence these brain functions, through the disposition of the key chemical molecules, along with the generation of action potentials in the sensory axons innervating the oral sensory receptors, which convey the patterned information from the oral senses to the cerebral cortex for decoding and sensorimotor integration function, an obvious circumstance that is relevant to the development of normal oral motor behavior. Conversely, sensory deprivation of the cerebral cortex may affect learning of oral functions, such as speech and chewing, through the impairment of the brain's respective mechanisms.

The implications of these relatively new discoveries are profound. The life plan of a neuron is not in its genes. It is also experience-driven. Accordingly, the growth and function of the nervous system which forms the substrate of the oral sensorimotor functions is an epigenetic process, responding to environmental experiences. This means that the oral functions, such as speech, chewing, etc, do not progress from some innate functional oral abilities, but are new functions that need to be learned actively by the child through the sensory experiences. In fact Schanberg [73] demonstrated that environmental factors such as tactile stimulation can influence genes related to growth and development of the body, through endocrine factors. This is interesting because the first sensory input in life comes from the sense of touch and pressure while still in womb. Infants and children are dependent on touch stimulation for normal growth and development and for building their brain. During the first year of life everything the baby picks up goes into the mouth and is learned through the mouth's touching.

In these perspectives the neuromuscular disorders associated with malocclusions of the teeth may be regarded as brain dysfunctions affecting the mouth's functions. Orthodontic changes in the occlusion of teeth and of the maxillofacial skeleton may improve the abnormal oral functions through neuroplasticity, which in turn, may be incorporated into the rules that govern the structure and function of the oral sensory and motor maps in the postcentral gyrus and precentral gyrus of the cerebral cortex as the brain changes. This implies that orthodontic therapy can alter the sensorimotor behavior of the mouth, which is key to the soft and hard tissue anatomy of the mouth. This view is in contrast to the conventional orthodontic concepts, which attribute to orthodontic therapy changes a minor peripheral influence, confined mainly to dentoalveolar structures.

This book may help to set the scene for future explorations in this field, which may help to elucidate the abnormal oral motor behavior in malocclusions of the teeth, delving into the underlying brain mechanisms. Our challenge is to discover how oral experience builds the structure and function of the brain.

Athens, Greece Margaritis Z. Pimenidis

Contents

Experience Changes the Brain

1

1.1 Introduction

The brain is the seat of feelings and behavior. Your brain creates your world—a radical statement about ordinary thinking. It is your brain that perceives and experiences. Everything begins and ends in the brain. How our brain works determines the very quality of our lives. Our brain patterns help us with our behavior, motor skills, work and being successful in our day-to-day lives. When behavior becomes abnormal, however, often there is something the matter with the patterns in the body's computer—the brain.

Unfortunately, there are many dental professionals who lack sophisticated information on how the brain actually works. They believe the oral behavior of their patients is primarily the result of environmental stress or conditioning, and do not consider the possibility that it may be based on abnormal brain physiology. I believe we need to understand the role of brain physiology along with other factors such as stress or conditioning before we can design successful oral orthodontic reconstruction treatments for people.

The prevailing concept in orthodontic practice holds that the sensory and motor functions or behavior of the mouth, which is intimately related with the development of the oral–facial structures, is controlled by the brain and, therefore, cannot be influenced by therapeutic measures [1, 206]. This concept is not a flimsy one. It is based on a century-old dogma that mammalian neurons do not divide after development, and they are preprogrammed to develop normally. That is, each infant neuron contains all the information necessary to live, develop, and be well. In other words, the entire plan, the blueprint for a neuron's life, is in its genes. The brain is a complex but a static, immutable, hard-wired switchboard that cannot be changed [2]. Accordingly, the dogma that the nervous system is static has prevented a meaningful analysis of the interplay between neuromuscular activity and skeletal morphogenesis, through which an understanding of the etiology of malocclusions of the teeth could emerge.

Recently, however, two new discoveries have revolutionized our views. These are: the structural changes of synapses by the firing experience of neurons, and the discovery of growth factors in the brain. These are discussed next.

M. Z. Pimenidis, *The Neurobiology of Orthodontics*,
DOI: 10.1007/978-3-642-00396-7_1,

1.2 Oral Experience Strengthens Neuron Synapses

Neurons are brain cells that are involved in communication of information. The strength of information processing performed by a neural circuit depends on the number of synapses between neurons [3]. For years scientists imagined that the electrical impulses and the chemical signals (neurotransmitters) that neurons use to communicate with each other, do not change the structure of the synapses. They thought that a synapse simply transmits the action potential signal passively, like a telephone, without itself changing [2].

Hebb [16] first proposed that synaptic knobs might be reorganized with learning. Since then there have been hundreds of studies examining brain tissue for synaptic reorganization following learning or following comparable physiological stimulation that might resemble learning-related events.

After years of experimental studies, however, researchers have discovered that the neuron synapses do indeed change with the firing experience of neurons. For example, it has been found that increased electrical stimulation for ten seconds of hippocampal memory neurons results in dramatic strengthening of synaptic communication and increased neuron responsiveness to neurotransmitter signal molecules (increase in the number of receptors projecting through the surface membrane of neurons to which the neurotransmitter is attached), that persists for about six hours after stimulation has stopped. These changes in the memory neurons alter brain function. Specifically the changes in the memory neurons of the hippocampus accelerate information processing through strengthening of the synapses of the neural pathway [17]. In another experiment rats were taught to navigate through a complicated maze while the electrical activity of the hippocampal memory neurons, especially critical for spacial memory, was monitored. As the rats learned to navigate the maze successfully, the hippocampal neurons fired with a characteristic unique pattern which was associated with learning of the particular task [59].

Many studies have provided evidences that chemical and structural modifications of dendrite spine synapses underlie much of the plastic changes in the brain in response to learning an experience [3–5]. Morphological changes of the dendrite spine are key to its plasticity following processing of information [3, 6, 7]. Electrical stimulation of neurons leads to the formation of new dendrite spines [3, 8, 9]. In this regard, dynamic changes of the actin cytoskeleton of the spine have been implicated in the formation of new spines [3, 10, 11]. The actin cytoskeleton of the dendrite spine plays a central role in synapse formation and maintenance [3, 12]. Processing of new information in neurons causes remodeling of actin cytoskeleton protein and modifies synapse stability, which may lead to formation of new synapses [13–15].

Thus, we now know that synaptic change clearly occurs with learning. Synapses are not unchanging hard-wired connections between neurons as previously assumed. Synapses continually strengthen, weaken, grow, and shrink throughout the life of the nervous system in response to experience. A number of new studies have further addressed the question whether these kinds of learning-related plastic changes are permanent, and the consensus is that they are not. The question that arises then is how learning-related information is permanently stored [2, 15]. Several ideas have emerged that are discussed next in Chap. 4.

Synaptic contacts take place in dendrite spines which are small sac-like organelles projecting from the dendritic trunk. Dendritic spines are more abundant in highly arborized cells, such as pyramidal neurons. The number of excitatory inputs can be linearly correlated with the number of spines present in the dendrite. Dendritic spines contain ribosomes and cytoskeletal structures, including actin and tubulin proteins. The dynamics of cytoskeletal structures is central to our contention that information can be processed and delivered to the synaptic function by changes in the cytoskeletal structures of neurons [3].

1.3 How Experience Strengthens Synapses

To figure out how neuronal connections change, we have to know something about the mechanisms of action. A neuron communicates with another neuron by sending a chemical signal across the synapse. Depolarization of the terminal axon by the action potential causes the release of a chemical substance (neurotransmitter) in small multimolecular "packets" from the nerve terminal into the intercellular space. The neurotransmitter diffuses to the next neuron, usually separated by a gap (synaptic cleft) about 20–30 nm deep, and there it binds with the receptor projected from the membrane of the postsynaptic neuron. This binding leads to permeability changes in the membrane of the postsynaptic neuron. Depending on the chemical structure of the neurotransmitter and the chemical nature of the receptor substances (sites), the induced permeability change of the membrane leads to either excitation, i.e., it leads to initiation of an action potential, or inhibition, preventing or depressing the ability of a neuron to discharge. On most neurons there are both inhibitory and excitatory synapses, the balance of their influence determining whether a neuron will or will not generate an action potential or whether its frequency of discharge will increase or decrease.

Interestingly, the basic mechanisms for excitation and inhibition are similar. In both types of chemical synapses, transmitter substance is secreted by nerve terminals, and in each the transmitter increases the permeability of the postsynaptic membrane for selected ions only. At excitatory synapses the transmitter increases the permeability to certain cations (Na^+ and K^+) whose movement inside the cell will reduce the membrane potential and cause current flow which sets up impulse propagation. At inhibitory synapses the transmitter increases permeability for small ions (K^+ and Cl^-) which keeps the membrane potential below threshold and thereby prevents impulse propagation.

Whether a neuron will cause excitation or inhibition in the next neuron in line cannot always be attributed to the chemistry of the transmitter alone. Differences in the properties of the receptor may be equally important. For example, the motor nerve terminals at skeletal muscles secrete acetylcholine which combines with specific receptors at the motor end-plate. This leads to a simultaneous flow of Na^+ and K^+ ions across the membrane in the end-plate, causing a depolarization and conducted impulses, which in turn lead to an activation of the contractile mechanism in the muscle. In the heart the same neurotransmitter, acetylcholine, liberated by the nerve endings of the vagus nerve, causes an increase in the permeability for K^+ alone, leading to inhibition.

Synaptic strengthening requires the participation of both neurons forming the synapse. Both the signal-sending presynaptic neuron and the signal-receiving postsynaptic neuron have to fire simultaneously in order to result in a stronger synapse. If the presynaptic neuron discharges and sends an electrical potential signal, but the receiving neuron does not respond, strengthening of the synapse does not occur. Or if the postsynaptic neuron discharges in the absence of presynaptic neuron firing, the synapses between the two neurons are not strengthened [2, 16]. Synapses do not strengthen willy-nilly because scattered neurons fire. Information must actually be transferred from sending to receiving neuron to ensure strengthening of the synapses. This requirement implies that the neural circuit must be functionally active to result in synaptic strengthening [17, 18].

Hebb [16] suggested that the more a synapse is used, the stronger and more efficient at transferring information it becomes. Use in turn is driven by experience. Thus, the same synapse may undergo graded weakening or strengthening according to its use, the type of stimulation and the availability of key chemical molecules, such as neurotransmitters [2].

The results of experiments by Bliss and Lomo [17] suggest that experiences strengthen synapses of only the specific pathways processing the experience; experiences do not affect synapses that are not activated by the action potentials. Thus specificity is built into the process. Oral experiences use specific circuits and synapses, thereby strengthening those very synapses and pathways, resulting in heightened input and processing of information in the cerebral cortex [3].

Hebb [16] speculated that learning and memory of experiences occurs at the synapses. He proposed that learning consisted of strengthening of synapses of a pathway. Strengthened synapses bind the neurons together into "ensembles." If individual neurons represent different features of a scene, for example, the ensemble recreates an image of the scene in its entirety, like the pieces of a puzzle, resulting in learning.

1.4 Developmental Agents of the Brain

The other revolutionary discovery is that growth factors which normally reside in the body, are also present in the brain [19, 20, 30]. The growth factors are critical brain signals that strengthen the neuron connections, the synapses, enhancing then the transmission of electrical impulses in the brain [21, 22]. Brain growth factors convert experience into strengthened neuronal synapses that mediate learning and memory [2, 30]. Entire families of growth factor molecules, including nerve growth factor (NGF), epidermal growth factor (EGF), fibroblast growth factor (FGF) and brain-derived neurotrophic factor (BDNF), have been found to be produced in the brain, and different populations of brain neurons have been found to depend on a certain growth factor for growth and survival [23, 24, 30]. The implications are profound. The brain can no longer be considered a static immutable, hard-wired structure, but is a vast reservoir of growth factors that direct growth, survival, and flexibility of neurons [19, 20].

For example, the presence of NGF in the environment of cholinergic brain neurons (which release acetylcholine) causes division of neurons and activation of survival genes, which ensure neuron survival. NGF deprivation inhibits neuronal division and activates suicide genes, which cause neuron death [25]. Similarly, adding NGF to cultured "Alzheimer neurons" has been shown to extend the life of neurons that normally die in patients with Alzheimer's disease [26, 27].

In addition, it has been suggested that BDNF in the brain governs everything from vision to thought to consciousness, sensation, and movement of muscles. BDNF factor has also been found to support survival of cholinergic neurons that degenerate in Alzheimer's disease, the dopamine neurons that degenerate in Parkinson's disease and the spinal motor neurons that die in Lou Gehring's disease. BDNF is most abundant in a key memory center, the hippocampus [30]. Accordingly, it has been suggested that the brain neurons are programmed for both survival and suicide. External chemical signals determine whether the survival program or the suicide program of neuron is turned on [28, 29].

The most important finding, however, is that NGF appears to cause growth of nerve fibers, which subsequently innervate their targets. For example, during development before any nerves have sent fibers to target cells, each individual target releases NGF like a radio transmitter sending forth waves in all directions. In response, distant spherical nerve cells extend fibers that grow towards regions of ever-higher concentrations of NGF, until the target cell is reached. Once the nerve fiber has reached and has innervated the target, it is assured of obtaining the NGF hormone and surviving. Then the NGF, which is synthesized by the target, is taken up by the innervating nerve terminal and transported back through the nerve process to the nerve cell body. This process is called retrograde transport because it goes from the nerve terminal to the cell body in the opposite direction to the action potential. When the NGF arrives in the cell body via the retrograde transport biological actions are initiated, namely division of neurons, survival of neurons, and formation of fibers and operational circuits in the targets. Thus, the neurons whose fibers succeed in reaching the target live as a result of contacting a source of NGF. This mechanism is part of morphogenetic cell death, which occurs during development and generates the sensory and motor maps in the somatosensory cortex [2, 204].

1.5 Families of Developmental Agents of the Brain

Experience accesses the neurons through developmental agents, which can be divided into three families [2]:

1. Neurotransmitters, for instance glutamate, acetylcholine, norepinephrine (noradrenaline), dopamine, serotonin etc., are chemical signals produced by the neurons that convey the action potential from one neuron to another across the synaptic gap [2]. Neurotransmitters also bind to surface membrane receptors of neurons and initiate intracellular biochemical changes in neurons that underlie sensation, perception, learning, memory, motor control of muscles and mentation [15].
2. Growth factors, for instance NGF, also act as keys that fit receptor locks on the surface of neurons. These keys and locks initiate changes in the biochemistry and electrical properties of neurons that involve the number (division of neurons), size and shape of

neuron, the growth of nerve fibers, the formation of synapses and the strengthening of synapses [31, 32].

3. Trophic factors support neuronal survival and alter the excitability of neurons by increasing the number of electrically conducting (ion) channels in the nerves. Also, the trophic factors increase the responsiveness of neurons to neurotransmitter signal molecules by increasing the number of receptors projecting through the surface membrane of neurons, to which the neurotransmitter is attached.

All these three families of molecules cause neuron growth. We are therefore prompted to focus on molecules and neurons as units of information processing [2].

The submicroscopic world of molecules possesses reasoning power, or computational power, that is thousands of times greater than at the neuronal level. For example, a single enzyme molecule performs one million chemical reactions per second. In contrast, whole neurons convey impulses in thousandths of a second. Thus, the enzyme wins. It conducts its activities a thousand times faster than the neuron [2]. At every level of organization, from molecule to neuron to neural system to networks of systems, the process of growth distinguishes the brain. This means that extant molecules are modified, new molecules form, molecules aggregate to create new cell structures and neurons new connections. These are only a few of the processes the brain uses to incorporate change. Destructive, or regressive, growth events are equally important. Removal of molecules, remodeling of cell structures, weakening of neuronal connections and even cell death are also used by the brain to encode information. Growth, then, is a strategy unique to life, enabling us to learn new experiences, remember, think, feel and be human [2].

This capacity for change is built into the brain at multiple levels, and is also termed "plasticity." Plasticity is innate; it obeys the rules of biology. But individual plastic processes can be activated by the environment. Plastic growth mechanisms draw on neurotransmitter, growth factor and trophic factor molecules. Neurons use neurotransmitter and trophic factor communication to accomplish a remarkable feat: the conversion of momentary environmental events into neuronal changes lasting months to years. This is how it works. Brief experiences generate action potentials in sensory neurons that release neurotransmitter signals. The neurotransmitter jumps the synaptic gap to electrically stimulate the next neuron in line. The neurotransmitter activates genes in the postsynaptic neuron that make trophic factor, and this factor strengthens the synaptic connection. Synaptic strengthening can then last for weeks, at least. Thus, plasticity allows brief experiences to create fleeting electrical impulses; impulses release chemical molecules, chemical molecules elicit growth, growth stores information. In this view, experience itself regulates the growth of the nervous system and the ability of neurons to communicate between one another through the strengthening of synapses [2].

Growth mechanisms allow the brain to capture time. Brief experiences change the structure of synapses, and the changes persist over time, creating memories. Memory mechanisms are built into the rules governing brain architecture. Changes in the environment translate into changes in the brain. Plasticity allows us to embody time. We are not doomed to live in an eternal present; we are beings with a history. And awareness of history, made possible by memory, permits learning. When learning and memory fail, we confront baffling and frightening diseases. For example, brain neurons that form memories die in patients with Alzheimer's disease destroying the patient's brain, mind, and humanity [2, 33].

1.6 Trophic Functions of Neurons – Plasticity

The term trophic functions of neurons refers to the influences that neurons exert on the structures they innervate. Although trophic influences are mediated through synaptic connections, their action is not necessarily related to electrical activity at the synapse. In other words, in addition to transmitting neural information, synapses transmit chemical influences to the tissues they innervate. Indeed there is evidence that the chemical influences appear to be transmitted in both directions, i.e., from the cell body to the nerve terminal and back. Specifically, Hendry [204] demonstrated that radioactive NGF injected into the target tissue is taken up by the innervating nerve terminal and transported back through the nerve process to the cell body where biological actions are initiated. Note that electrical impulses are propagated in the forward direction, from the cell body to target terminals, while NGF travels by retrograde transport in the opposite direction. These processes (neurotrophic influences) could indeed serve as effective regulators in the embryological development of the nervous system. The trophic substances could provide the types of chemoaffinities necessary for the development of highly specific connections. For example, it has been postulated that trophic substances traveling from motor neuron to muscle and released by the nerve terminal could determine the size of the chemosensitive area of the muscle membrane, as well as the contractile properties of the muscle [203].

The neurotrophic influences strongly support the notion that these processes are essential for the development of appropriate connections in embryogenesis, and for the maintenance and regulation of these connections in later life as well. If the neural connections are largely specified genetically, how do functional modifications occur in the brain? The environment continually produces modifications in the behavior of organisms through learning. The study of learning begins with the fact that a variety of neuronal networks undergo persistent changes which are analogous to information storage. Thus, plasticity of neurons expresses the unity of our internal and external worlds, the unity of organism and environment, the unity of nature and nurture. In this context, the Promethean insight that the environment regulates the brain which regulates the environment is emancipating. It provides the therapeutic keys to free us from the devastating disorders of mind. Genes, molecules, cells, systems and brain regions interact with the environment to generate cognition, consciousness, motor control and emotions, as described in Chap. 4.

1.7 Oral-Cavity – Reservoir of Growth Factors

A number of brain-derived growth factor molecules are also produced in the salivary glands participating at least in glandular development [201] and their secretion in the oral cavity is increased with mastication [200]. Specifically, Levi-Montalcini [205] discovered a hormone in the mouse salivary glands that induces in vitro growth of the sensory and sympathetic neurons. The hormone was termed NGF because it prompted neurons to grow in size and number, and to send out long fibers. These are fibers that neurons use to contact one other. These fibers develop specialized contacts to form synapses. Synapses transmit information from one neuron to another, which underlies information for learning and memory. Thus, the NGF of salivary glands might be a potential link between oral experience and the brain.

Levi-Montalcini treated live rats and mice with NGF and saw similar effects. Again NGF treatment increased the size and number of neurons lying outside the brain. It also caused a stunning increase in the number of nerve fibers in target organs of the neurons, such as the heart and iris. NGF appeared to be increasing the connection of nerves and their targets in living animals.

Since NGF has such dramatic actions in the animal body, it was natural for Levi-Montalcini to ask what is the reaction of the body to NGF deprivation. She administered an NGF antiserum which could neutralize any NGF present in the animals' body and prevent actions on neurons. The results were again dramatic. The NGF-neutralizing antiserum caused the death of many neurons and essentially destroyed all the fiber connections of the neurons with their target organs. The NGF antiserum experiments suggest that NGF is in fact normally present in the body, produced at least in the salivary glands, and that its presence is necessary for neuron survival. The most important finding, however, was that NGF appears to cause nerve fiber growth and connection of nerves with their targets.

EGF is another developmental agent of the brain, as well as of the epithelial tissues of the body. It is a small protein found in the submandibular and parotid glands from which secretion in the oral cavity is increased with mastication [200]. During mastication afferent inputs carry sensory information mainly from the periodontal mechanoreceptors to the masticatory muscles, as well as sensory signals from taste-activated chemoreceptors in the taste buds, to the salivary nuclei in the medulla. The efferent arm of these masticatory and gustatory stimuli is the autonomic nervous supply of the salivary glands. The classical neurotransmitters released from the peripheral postganglionic parasympathetic and sympathetic nerve endings are acetylcholine and noradrenaline (norepinephrine), respectively. These molecules bind to specific receptors projecting from the membrane of the salivary gland cells and alter its permeability to ions, thereby initiating biological actions inside the cell, which determine the rate of flow and the composition of saliva in the oral cavity, in response to oral sensory experiences generated by the masticatory movements.

Another growth factor, which was originally found in the peripheral tissues of the body and subsequently in the brain, and which acts as a developmental agent of neurons, is FGF. It is also a potent regulator of wound healing and is found in saliva [200]. FGF stimulates the growth of fibroblast cells of connective tissue to secrete tropocollagen molecules which by condensation form the collagenous fibers, and thus participates in salivary gland development [201].

Accordingly, it is hypothesized that the FGF might also control collagen formation in the oral cavity and the masticatory muscles. The connective tissue derives from neural crest cells in the first branchial arch into which the myoblast cells migrate. This may suggest that FGF is actively involved in oral morphogenesis. A similar hypothesis is that of Copenhaver [202] who suggested the existence of a "regional factor" in the stomodeal ectoderm, which influences the invagination of the mouth and the formation of the oral cavity. It is possible then that FGF and/or EGF participate in oral morphogenesis through an induction interaction phenomenon. Kallman and Grobstein [201] found that during the induction interaction of oral epithelium with the mesenchyme, collagen deposition at the surface of epithelium is important for salivary gland development.

In this context, it can be speculated that the embryonic oral region is a reservoir of various growth factor molecules produced by target cells that ensure the growth and development of tissues and organs of the mouth, which in turn exert a growth modifying effect on the nervous system which innervates them (nerve modulation). For example, Weiss [203] suggested that the parotid gland will impart its specificity to the autonomic nerve fiber to which it is attached. Similarly, the motor nerves of muscles and the sensory nerves are modulated by their respective end-organs.

It has been suggested that the amount of NGF produced by the target brain cells controls target size. Target size in turn controls how many nerves grow into the target, how many nerves are modulated through retrograde transport, and how many neural pathways and hence lines of communication are formed between the mouth and brain in order to conform with normal oral behavior [2, 204]. In other words, how the blueprint for information input and processing in the brain is organized, linking the oral experience/activity with the motor control of the masticatory muscles through the sensorimotor integration functions of the cerebral cortex. In addition, the presence of growth factors in the oral cavity with morphogenetic influences may also suggest that oral morphogenesis is an epigenetic event and the growth factors are central to this development.

1.8 The Ever-Growing Brain

Surprising relationships among growth factors, connectivity between neurons, neural communication and brain function, as well as oral development are now emerging. Growth factors are critical brain signals that regulate survival or death of neurons [2, 34]. Growth factors are chemical signals that strengthen the neural connections (synapses) lying at the heart of brain function. Brain growth factors convert oral experience into strengthened neuronal synapses that mediate communication between neurons and accelerate information processing in the cerebral cortex that underlies sensation, perception, learning, memory, motor activity, and mentation [2, 15]. However, the most outstanding recent surprise in neuroscience is that contrary to traditional teaching, growth of the brain is not governed by internal programs. Growth is also experience-driven. An adult brain can make new neurons through experience. As observed in mice, new neurons in the memory-enhancing hippocampus especially bloom when mice take to the running wheel [23]. Thus, the elixir for the brain, which appears to be the best natural cardiovascular therapy of all, is exercise [35].

The implications are profound. The brain is a growing ever-changing flexible structure that generates our plastic minds in response to experiences which generate the action potentials in the nerves, which in turn triggers the neurons to produce chemical signal molecules. In principle this implies that since the neural impulses and chemical signal molecules are responsive to experience, the distinction between nature and nurture breaks down. A neuron is not enslaved by its genes. There is not enough information in each neuron's genes to direct development in the brain. Additional instructions come from the world outside the neuron as growth and survival signal molecules [23, 24]. Contrary to what was believed before, the brain neurons are renewable resources. New brain neurons can indeed be produced in the adult brain through a vast reservoir of growth factor molecules [2].

Black [2] gives the following list to gain perspective of neuroscience discoveries that have revolutionized our approaches to brain functions and diseases.

- Specific populations of neurons in the brain control specific behaviors and discrete mental functions.
- The brain is home to myriad growth and survival chemical factors that govern the life and function of specific cell populations.
- Identified, cloned genes control selected mental abilities and behaviors.
- Specific environmental stimuli, whether physical, verbal, or situational, activate specific brain genes, expressing plasticity.
- Contrary to traditional dogma that the brain neurons do not divide, new neurons can actually be produced in the adult brain from "stem cells." This means that the brain may be able to replace its own degenerating neurons.
- An enriched environment and physical activity increase proliferation of brain cells.
- Genes for pain signals and receptors have been cloned. Thus, with an emerging understanding of the physical basis of pain, we are now able to eliminate major forms of pain. We are rapidly approaching a time when we can banish pain itself.

These discoveries comprise a blueprint for developing new treatments for formerly hopeless neurological diseases, such as Alzheimer's, Huntington's, Parkinson's, and Lou Gehring's diseases. In addition, growth factors and inhibitory factors have been discovered in the spinal cord, and paralyzed rats may recover through regrowth of spinal nerves.

All these discoveries though exciting and revolutionary, involve abstractions that lie beyond visualization. Neuroscience, however, provides the solution of this problem. Functional magnetic resonance imaging (fMRI) and positron emission tomography (PET) can picture the brain systems underlying specific thoughts, learning and memory of experiences, nerve-muscle diseases, emotions etc. We can also see brain dysfunction in oral neuromuscular disorders associated with malocclusions of the teeth, as discussed in Chap. 10.

Functional and Dysfunctional Aspects of the Cerebral Cortex

2

2.1 Introduction

Diagnosis of oral motor disorders is a professional skill. Most orthodontists and other specialist dentists are familiar with the many types of oral motor behaviors and their dysfunctions, such as normal chewing, speech, improper bites, malocclusions of the teeth, and oral–facial imbalances, but have perhaps not thought too much about the underlying processes or mechanisms that regulate these behaviors and which may eventually provide practitioners with a rationale for correcting dysfunction. Thus, special interest is focused on the information–input–processing function of the brain that is necessary for giving a conscious experience and how it may be related to oral motor behaviors that underlie the structure of the mouth.

In this context, the basic goals of oral practitioners are to discover the patient's motor problem, understand the potential causes of the problem, the ways the central nervous system regulates motor behavior, and to recommend appropriate treatment. To accomplish these goals the diagnostician must have a basic knowledge of normal and disordered oral motor behavior. Since the clinicians must mediate between the patient and the dysfunction when providing corrective experience it is important that they can conceptualize the nature of oral motor behavior.

The information processing approach to motor control and learning provides a conceptual framework in which to study and understand oral motor behavior, for instance during chewing and swallowing, within the control of a mouth system with ever-changing behavior.

2.2 The Sensorimotor Cortex

The somesthetic cortex or neocortex (excluding the limbic cortex) integrates information received from the receptors throughout the body, including the mouth. The somesthetic cortex (postcentral gyrus, Brodmann's areas 1, 2, and 3) is organized vertically. Columns of neurons (module), from one to several neurons, are oriented vertically to the cortical surface. Many vertical cylinders of neurons constitute this area. The column of neurons or module in

M. Z. Pimenidis, *The Neurobiology of Orthodontics*,
DOI: 10.1007/978-3-642-00396-7_2, © Springer Verlag Berlin Heidelberg 2009

the somesthetic cortex of the postcentral gyrus is the basic neuron circuit that in its elemental form comprises input channels (afferent fibers), complex neuronal interactions in the module, and output channels. The output channels are largely the axons of the pyramidal cells of the motor cortex of the precentral gyrus, which directly or indirectly through the pyramidal and extrapyramidal tracks, respectively, innervate the alpha motor neurons of skeletal muscles, including the trigeminal system. The postcentral gyrus, however, also has a role in motor activity, since the electrical stimulation of the gyrus may elicit movement at times. Similarly, the precentral gyrus is also a "primary sensory cortex" because it also receives input from the ventral posterior thalamic nucleus as does the postcentral gyrus [36].

The cerebral cortex is also organized into six layers or laminae of cells, arranged parallel to the surface of the brain, each of which contains different cell types that play different roles. Beginning at the cortical surface the layers are: (1) the plexiform layer or molecular layer, (2) the external granular layer or layer of small pyramidal cells, (3) the layer of medium-sized and large pyramidal cells or external pyramidal layer, (4) the internal granular layer or layer of small stellate and pyramidal cells, (5) the inner or deep layer of large pyramidal cells, and (6) the spindle cell layer or layers of fusiform cells or multiform layer [36].

Pyramidal cells of layer III project to other cortical regions (corticocortical connections), whereas pyramidal cells of layer V project to subcortical structures conveying motor commands to muscles. Layer V pyramidal cells are the largest. Layer III pyramidal cells are generally smaller. Nonpyramidal local circuit cells, which do not project out of the immediate cortical area, are smaller and more numerous than pyramidal cells. Some of these local circuit cells can be excitatory local relays, and others mediate local inhibition. One type of local circuit cell, the horizontal cell, interconnects cells across the module and is generally inhibitory. Minicolumns and modules are organized around the large pyramidal cells. The many excitatory local circuit cells relay information up and down the vertical column. In this manner, the large dendrite tree of pyramidal cells receives many inputs via local circuit cells [15].

Pyramidal cells receive many of their inputs onto dendrite spines. Spines are appendages of dendrites that increase their surface area, and compartmentalize the molecular machinery relevant to receiving inputs. As the cortical hierarchy increases, pyramidal cells receive more inputs. For example, pyramidal cells in the prefrontal cortex have 23 times more spines (and inputs) than the pyramidal cells in the primary visual cortex [15].

2.3 Oral Sensorimotor Integration

Oral sensory integration is a child's ability to feel, understand, and organize sensory information from his/her mouth. Sensations flow into the brain through oral experiences, and the brain locates, sorts, and orders sensory stimuli into sensations. When the stimuli flow in a well-organized or integrated manner (patterned input) the brain can use those sensory data to integrate functions underlying sensation, perception, learning, memory, motor control of muscles and mentation. The connection between the capacity of an individual to integrate sensory data and the child's social and emotional development should not be underestimated. How a child integrates information through the sensory system provides a basis of the child's oral and body reality. When the flow of sensory stimuli is disorganized, life can be like a rush hour traffic jam. It is sensory integration that attempts to "put it all together" and that helps us make sense of who we are and understand the world around [78, 92].

The integration of oral sensory and motor functions or behavior depends on the development of reflex circuits connecting the sense organs through the central nervous system to muscles. The sensorimotor integration function of the mouth can then be analyzed in terms of three basic functions:

1. An input transducer function. This includes all the sophisticated sensory receptors and their innervating sensory neurons, which are endowed with specificity, that is to say are able to transform definite kinds of external stimuli into complex electrical and chemical phenomena, depending on the nature and strength of the stimulus, and convey patterned action potentials (coded information) from the senses to the somesthetic cortex.
2. A control function carried through the central processing unit, the cerebral cortex, namely the postcentral and precentral gyri, which represent the primary somesthetic and motor cortex. The capacity to distinguish one sensory experience from another in the cortex is known as discrimination. Specifically, the cortical neurons must be capable of decoding the frequency of the action potential of each nerve transmitted to the brain in order to interpret the sensory messages of the receptors, and to store in memory information of the learned particular sensorimotor reflex experience.
3. An output transducer function. The human brain processes information in countless ways, most of which are poorly understood, if at all. When, however, there is stimulation of the oral motor areas in the cortex, movement of the mandible and tongue may result as a result of activation of trigeminal and hypoglossal motor nuclei, respectively [92].

By definition these three basic functions with relationships among them form a system, the oral sensorimotor integration system. There is no direct link between the oral input and output in the system. This means that the oral sensorimotor integration reflex function is an open loop. The motor output, however, has an influence on the sensory input. That is the movements of the mouth stimulate the sensory receptor input that guides oral function. Thus, the mouth's sensory experiences are generated principally by its own actions, and its actions are responsive to its sensory experiences [43, 65]. This means that the oral sensory system is functionally integrated with the oral musculature.

Similarly the receptors of the skin of the oral–facial region are functionally integrated with the underlying muscles of facial expression. The contraction of muscles stimulates the receptors through deformation of the skin [78, 93]. Granit [44] suggested that adjacent portions of the stimulated skin may be inhibitory to muscle action. This function may be locally regulated without reaching conscious states in the cortex. Thus, the skin has also established "local signs" for reflex action, and not only for conscious perception. This may imply that in certain circumstances the movements of the mouth may be initiated locally without reaching perception in the cerebral cortex. In this view, the oral mucosa, like the skin, may be able to deliver both general and localized messages. The general piece of information may be based on very large receptive fields, in which the afferent fibers overlap. The localized messages may be based on smaller receptive fields down to punctiform, which are innervated by a single afferent fiber and from which a certain type of reflex response could be elicited. The central nervous system may correspondingly be organized to take care of large receptive fields, reaching the conscious level, and others of small receptive fields for local function at the brainstem level [44].

The strength of information processing performed by a cortical circuit depends on the number of interneuronal connections or synapses. Morphologically speaking, a typical neuron presents three functional domains. These are the cell body or "soma" containing the nucleus and all major cytoplasmic organelles, a single axon, which extends away from the soma and has cable-like properties, and a variable number of dendrites, varying in shape and complexity, which emanate and ramify from the soma. The axonal terminal region, which connects via synapses to other neurons, displays a wide range of morphological specializations depending on the target area [3].

Turning morphology into a functional event, the soma of neurons and the dendritic tree are the major domains of receptive (synaptic) inputs. Thus both the dendritic arborization and the axonal ramifications confer a high level of subcellular specificity in the localization of particular synaptic contacts on a given neuron. The extent to which a neuron may be interconnected largely depends on the three-dimensional spreading of the dendritic tree [3].

Dendrites are the principal element responsible both for synaptic integration and for the changes in synaptic strengths that take place as a function of neuronal activity [46, 47]. While dendrites are the principal sites for excitatory synaptic input, little is known about their function. The activity patterns inherent to the dendritic tree-like structures, such as in integration of synaptic inputs, are likely based on the wide diversity of shapes and sizes of the dendritic arborizations [48]. The size and complexity of dendritic trees increase during development [49], which has been, in turn, associated with the ability of the neural system to organize and process information, such as when animals are reared in a complex sensory environment [50]. Thus, dendritic size and branching patterns are important features of normal brain development and function.

Most synapses, either excitatory or inhibitory, terminate on dendrites, so it has long been assumed that dendrites somehow integrate the numerous inputs to produce single electrical outputs [3]. It is increasingly clear that the morphological functional properties of dendrites are central to their integrative function [51, 52]. Branching in this context is essential to the increasingly complex ability of the neuron to respond to developmental complexity [48]. The dendritic cytoskeleton plays a central role in the process of ramification, filopodia formation, and more specialized neuronal activity such as long-term potentiation circuit dynamics, including memory formation, and the response to pathological conditions [3].

Synaptic contacts take place in dendritic spines, which are small sac-like organelles projecting from the dendritic trunk. Dendritic spines are more abundant in highly arborized cells such as pyramidal neurons and scarcer in smaller dendritic tree interneurons with few interconnections. The number of excitatory inputs can be linearly correlated with the number of spines present in the dendrite [3]. How information is processed in a dendritic spine, is still a matter of current study [6]. Dendritic spines contain ribosomes and cytoskeletal structures, including actin, and a- and b-tubulin protein. As in other parts of the neuronal structure, cytoskeletal components are highly relevant in the structure and function of the neuron. The axonal hillock contains large "parallel" bundles of microtubules. Axons are also located with other cytoskeletal structures, such as neurofilaments, which are more abundant in axons than in dendrites. The axon hillock is the region where the action potential is generated. The dynamics in cytoskeletal structures are central to our contention that information can be processed and "delivered" to the synaptic function by changes in the cytoskeletal structures [3].

The recently advanced new learning and memory model of Penrose and Hameroff [113] suggests that information processing and storage of memory occurs in the microtubules and microtubule-associated proteins of the subsynaptic zone of dendrites, which gives rise to the spine. Microtubules in the subsynaptic zone are capable of storing important information concerning where spines were originally located, and of using that information to determine where synapses will reappear following learning or enriched experiences [39–41].

The primary receptive area of the cortex integrates the more complex aspects of touch, deep sensibility, pain, and temperature. These perceptions include the appreciation of the location and position of a body part (proprioception), the localization of the source of pain, temperature, and tactile stimuli, as well as the comparison of these sensed modalities with those formerly experienced through memory [36]. Ablation of the postcentral gyrus is followed by loss of the finer and more subtle aspects of sensory awareness. For example, when an object is handled with the eyes closed, the subject can feel it but cannot appreciate its texture, estimate its weight or temperature. Difficulty is also experienced in appreciating the position of the body or body parts in space [36].

The sense of taste is claimed to be located at the base of the precentral and postcentral gyri, slightly rostral to the primary somesthetic cortex or in the cortex immediately adjoining the insula area [42].

In sum, as we view the development and function of the mature mouth we are more aware that the mouth is essentially an information-processing system, in which the higher faculties of the brain, namely learning, memory, motor control of muscles and reasoning, are chiefly concerned with the choices involved in getting food, and the development of the computational activities of mastication, speech, taste, etc, through learning and memory of the oral sensorimotor integration function of the cerebral cortex. Any animal would starve to death if the effect of food was limited to its action on the surface of the mouth. Thus, the enormous significance of feeding acquired by experience and learning becomes evident.

2.4 Sensory Integration Dysfunction

All the basic five sensory systems need to be working simultaneously and cooperatively for acquisition and performance of higher skills. In fact, if any one system does not work properly either by itself or in conjunction with the others, a sensory dysfunction of some kind may result. The degree of sensory dysfunction, whether it manifests as merely a "sensory issue" or mild, moderate, or severe sensory dysfunction, is highly dependent on which senses and/or sensory systems are impaired and to what extent. Again, the sensory systems actually operate in conjunction and cooperation with one another, so the resultant behavior or sensation is probably the outcome of accessing more than one system [79].

Often a child with sensory integration dysfunction will be categorized as immature and possible spoilt. The new social, cognitive, and motor demands of the school setting often create even more confusion, anxiety, and chaos for the child who has sensory integration dysfunction [79]. Now, let us explore some of the systems that may indicate sensory integration dysfunction.

2.4.1 Dysfunction in the Proprioceptive System

This refers to an inefficiency in the proprioceptive system that may cause a child to have difficulties both at school and at home. Since the proprioceptive system gives feedback from muscles and joints, it supports holding a pen or a pencil correctly, staying properly seated in a chair, and learning to use a knife, fork and spoon correctly. This system also supports learning to walk, opening and closing a jar or a door, using playground equipment, and chewing [79].

2.4.2 Dysfunction in the Tactile System

A dysfunction in the tactile system may manifest in a variety of ways. A child may be actively defensive and not want to be touched, bumped, or hugged, or to touch certain textures or wear certain types of clothes. An atypical tactile system may also cause a child to crave certain tactile sensations and experiences, such as never being able to keep their hands to themselves, constantly touching and feeling the people and/or things around them. In other words, the tactile system greatly influences relationships, fashion sense (clothing choice), and school performance [79].

2.4.3 Dysfunction in the Vestibular System

A problem or inefficiency in the vestibular system may cause difficulty with balance, coordination and motor planning. This may manifest in the child as clumsiness and uncoordinated fine motor activities, such as eating, drinking, using writing utensils etc. The vestibular system will play a key role in the ability of the child to participate successfully in sport, in the gym class or on the playground, and in various activities at school and at home [79].

2.4.4 Possible Sensory Signs and Symptoms

Let us now explore a wide range of behaviors and symptoms that may indicate some level of sensory dysfunction. This list is not meant to be used for diagnostic purposes, but rather to create awareness of some of the sensory difficulties a child may experience over the course of development and for which specialist help might be needed.

- Young children may be overly sensitive to touch. Touch is interpreted as a nociceptive stimulus and therefore working in their mouth is very difficult. For the same reason these children dislike bathing, brushing their teeth and other self-care activities. These children may be oversensitive to pain.
- Young children may be underactive to touch. These children exhibit oral hyperactivity by constantly eating or chewing. These children may be "chewers"—they chew shirts, blankets, toys, toothpaste etc.
- Children who do not tolerate the taste of certain foods, and therefore vomit or gag easily, and always have the same lunch; i.e., they are very picky eaters. This suggests gustatory sensory dysfunction.
- Children who do not know when their mouth is full and so stuff their meal into the mouth. This behavior suggests proprioceptive, tactile sensory dysfunction.

Traditionally, sensory integration has been the exclusive realm of the occupational therapist. Now this is starting to change and, although it is still the occupational therapist who will do much of the formal, standardized testing for sensory dysfunction, other specialists are beginning to recognize sensory integration dysfunction, including orthodontists, make referrals for further evaluations, and incorporate a "sensory approach."

2.5 Oral Experience Changes the Brain

After centuries of debate about body and soul and the nature of man, we now understand that the brain's architecture gives rise to mind, emotions, and personality. The tragedy of Alzheimer's disease and dementia, which are accompanied by personality fragmentation, bear undeniable witness that the brain is the basis of mind. Neuroscientists widely accept that cognition and consciousness are correlated with the physiological behavior of the material brain, and that the matter that comprises brain gives rise to its functions, in particular higher cognition and consciousness, through integration with the environment [33].

Sensory systems, such as the mouth, transmit sensory information to the brain through the sensory pathways. An oral sensory pathway begins in a sensory organ, usually makes synaptic contact in the thalamus, and then relays to layer IV of the primary somesthetic cortex through a series of corticocortical circuits. Change in the brain begins when the new experience activates the oral sensory pathway conveying information to the brain. These records are established through growth changes in the pathways and in the cerebral cortex [15].

The initial response of the cortex to a peripheral stimulus is the evoked potential. Immediately afterwards there is a change in the frequency of firing in the module, resulting in the signal becoming more sharply defined by elimination through inhibition of all the weaker excitatory stimuli. As a consequence the stimulus can be more precisely located and evaluated by the cortex. In fact inhibition that sharpens the neuronal signal occurs at each relay station in the pathway. Each relay station also gives the opportunity to modify the coding of the messages transmitted from the receptor organs [37].

Neurons in the pathway generally communicate with each other using chemical messages called neurotransmitters. Glutamate is the main excitatory neurotransmitter, which transmits sensory information along the sensory pathways throughout the central nervous system. Glutamate is produced in the signal-transmitting presynaptic neuron by proteins related to synthesizing, packing and releasing the neurotransmitter from the presynaptic nerve terminal. Glutamate traverses the synaptic gap and alters the membrane potential of the signal-receiving postsynaptic neuron. If the signal is strong enough, or if it is accompanied by sufficient additional input, the receiving neuron will generate an action potential that will reach another neuron in line. Thus, the sequence of events repeats itself [15].

Glutamate mediates excitatory sensory input to the cerebral cortex where many of these presynaptic terminals contact dendrite spines by binding to NMDA, AMPA, or kainate receptors located on the surface membrane of the spines. For example, glutamate binds to the NMDA receptor within few milliseconds, altering the ion conductance across the membrane. Electrically charged calcium ions (Ca^{2+}) enter the neuron and stimulate calcium-activated proteins inside the cell. The activated proteins initiate polymerization and depolymerization cycles leading to breakdown and then rebuilding of the microtubule protein, tubulin, as well as of microtubule-associated protein 2. These changes occur in the subsynaptic zone of dendrites lying beneath the spine which gives rise to the spine [15]. Thus, NMDA receptors mediate breakdown and then rebuilding of glutamate synaptic sites on the spines in response to information input, for instance, of oral experience. In this view, oral sensory information changes the conformational state of neuron cytoskeletal proteins, i.e., changes the brain through learning and memory of new experiences. Evidence suggests that memory is hard-wired in dendritic cytoskeletal structures [54–56]. Accordingly, the NMDA receptor is key to brain neuroplasticity [33].

Similarly, glutamate binds to AMPA receptors and opens sodium (Na^+) channels across the membrane. Sodium enters the neuron, increasing the sodium concentration inside the cell. This is likely to initiate polymerization of tubulin, the building protein block of microtubules. If the sodium concentration becomes too high, however, tubulin depolymerization will occur [15].

Thus, brief oral sensory experiences, for instance the taste of a substance or the exploration of an object in the mouth, generate fleeting electrical impulses in the neurons that innervate the stimulated receptors. The impulses make the transmitting neurons release molecular signal keys that selectively interact with specific receptor locks on the surface of the target neurons, eliciting biochemical changes and biological responses in targets underlying sensation–perception, learning–memory, and motor activity depending on the stimulus [15, 38, 58].

The signal from taste may also reach through local cortical circuits the hippocampus memory center and excite large specialized groups of neurons there that fire in a characteristic synchronized pattern that represents memory of learned experience [17, 57]. The memory mechanisms are then built into the rules governing the architecture of the cerebral cortex, in other words are built as changes in brain structure [38, 58]. Subsequently, the hippocampus sends a tract to hypothalamus that regulates the consumatory action pattern of the individual.

Similarly, when we chew food or gum the oral information is integrated in the sensorimotor cortex. The motor cortex then sends action potentials down to the alpha motor neurons of the trigeminal nuclei, through the pyramidal and extrapyramidal pathways, which in turn make the masticatory muscles contract. The transmitting neuron releases neurotransmitter signal, acetylcholine, at the relay station, which stimulates electrically the receiving neuron. Acetylcholine binds to specific receptors of the receiving neuron and strengthens the synapse. In this way, the strengthening of synapses through the signal molecules allows the neurons to hold electrical conversations for enough time to evoke, for instance, a muscle contraction or a taste experience [2].

Thus, molecular signals (glutamate, acetylcholine, dopamine, etc – hundreds of them) govern the function of the brain and mind, and increase the ability of neurons to communicate with each other through the strengthening of synapses. Not surprisingly then brain signals and receptors play a critical role in a variety of dysfunctions and diseases and treatment procedures. Signal derangement results in a wide variety of neuropsychiatric diseases including depression, dementia, and Parkinson's and Alzheimer's diseases [2].

Fascinating experiments have provided insights into how experience induces brain changes. For example, rats were taught to navigate through a complicated maze, while the electrical activity of their hippocampal memory neurons was monitored. As the rats learned to navigate the maze successfully, the neurons fired with a characteristic (unique) synchronized pattern that represented memory of the learned task [59].

In another experiment, maze-learning rats were stressed with a mild electric shock to the tail. A control group received no electric shock. The group that received the shock stressor learned faster and retained the knowledge longer than the control group. Moreover, in the stressed rats neuronal connections were strengthened more efficiently than in the control group [60, 61]. Accordingly, it has been suggested that some degree of stress may promote some forms of learning, through alertness and attention of brain states that foster synapse strengthening and learning. In this view, neuroscience may provide new strategies for learning and teaching in school. Memory abilities are subject to change and hence and a little stress may sharpen memory through strengthened synapses [2].

Hebb [16] suggested that synapses in the brain are strengthened in accordance with how often they are used (stimulated), enhancing then the input and the processing of information, which in turn contributes to the richness of sensory experiences. Hebb speculated that strengthened synapses bind the neurons together into "ensembles." If individual neurons represent different features of a scene, for example, the ensemble recreates an image of the scene in its entirety, like the pieces of a puzzle. In this view, strengthened synapses may associate for instance, visual, auditory, emotional and conceptual features of a memory that are encoded in the component neurons.

In addition, experimental studies by Bliss and Lomo [17] suggested that experiences strengthen synapses of only the specific pathways processing an experience. Experiences do not affect synapses that are not activated by the action potential. Specificity is then built into the process of transmitting information; the information is transmitted through activated (specific) pathways. Accordingly, learning and memory are also specific processes, since they use specific pathways.

Thus, the communicative connection between neurons, the synapse, and its strengthening by electrical impulses and chemical signal molecules through use, is critical in regulating oral information input and motor output. If the synapses in the activated pathways are strong more information is relayed and processed in the cortex, resulting in learning, memory, perception, and motor control of muscles. On the other hand, if the synapses are weak, less information is communicated [2].

Neurophysiological studies indicate that the sensory nerves are not high fidelity recorders of the peripheral stimulus, because they accentuate certain stimulus features and neglect others. In addition, the central nervous system, which does the processing of the information, is never completely trustworthy, allowing distortions of quality and measures. Because of this there is general agreement that sensation is an abstraction, not a replication, of the real world. The more specific is the sensory stimulus perceived by the brain, the more specific is the motor pattern [62].

While the central nervous system needs continuous stimulation of all sensory modalities during childhood to maintain the strength of synapses in order to increase the input and processing of information, which is the key to sensation, perception learning, memory and motor control of the mouth, on the contrary idleness of the mouth decreases the strength of synapses and inactivates the sensory pathways to the brain. Hence, very little information is entering the brain and the cerebral cortex is suffering from sensory deprivation [63]. Under these conditions the cortex will reduce the input to the motor cortex controlling the motion of the mouth and the voluntary masticatory movements, leading to oral dysfunction and to delayed maturation of oral functions [64].

The neural events pertaining to the normal or abnormal function of the mouth are registered in the cerebral cortex at all stages of the development of the mouth. This ability of the brain for self-regulation of its structure and function, according to oral sensory information-processing and learning–memory capacity, is called plasticity of the brain, and is reflected point-to-point in the anatomical representation of the mouth in the sensory and motor maps of the cerebral cortex. The cerebral maps are continually modulated by the various sensorimotor functions of the mouth reflecting experience-driven changes in the brain [43, 65]. Thus, the normal oral sensorimotor functions, such as speech, chewing, swallowing, taste, etc, will be seriously disturbed without normal sensation and perception of oral structures, without normal jaw and tongue movements, and without adequate secretory activity of the salivary glands [66].

In this view, mastication of food might be an important motor activity, contributing to significant sensory input and to structural and functional development of the brain. Similarly, any dental or orthodontic procedure that restricts the movements of the mouth for a long time may affect the normal sensory input and the integration of sensorimotor functions in the cortex.

Kawamura [66] suggested that both life experience and emotion contribute to individual variation in oral sensorimotor functions and in particular to the masticatory function. Bosma [67] suggested that the infant organizes and governs its sensorimotor world of experience in relation to its own current needs and purposes. In doing so the infant is generating its own neurological future. This reaction of the infant to external influences is not a passive one. Instead the child is striving toward the outer world, and this is an instinctive involuntary phenomenon. When this striving is satisfied, i.e., when it causes, for instance, a movement, the movement is a reflex in the full sense of the term. There is no doubt that complete dependence on this instinctive striving is responsible for the extreme mobility of infants and children, which constantly pass from exercise of one nerve to that of another. It is then this mobility that ensures the normal development of the sense organs, including the oral senses, resulting in organized motor behavior. Thus, the first condition for maintaining the structural and functional integrity of nerves and muscles is adequate exercise of these organs [68].

2.6 Role of Reticular Formation in Oral Sensorimotor Functions

The reticular formation is believed to regulate or control the sensory input to the brain. The feedback information supplied between the sensory input and motor output must be of the right amount if the operation of the sensorimotor function of the mouth is to be regulated for optimum results. Too much feedback or central modification of the sensory input may mean sensory overload for the brain, which can be catastrophic leading to wild oscillation in oral motor behavior. Conversely, reduced sensory input may mean sensory deprivation of the brain, again deranging oral motor behavior. Thus sensory overload, sensory deprivation and the central disposition of the individual may reduce the arousal reaction of the cerebral cortex for attention, awareness and learning, leading to unconscious reduction of interaction of the individual's mouth with the external environment through changes in the reticular formation and the ascending reticular activating system (ARAS) [63, 64].

There are four major neurophysiological mechanisms dealing with the organization and functioning of the central nervous system, all related to reticular formation; these are discussed in the following sections.

2.6.1 Descending Reticular Control

The descending reticular control mechanism has been implicated in the facilitation and inhibitory roles of the reticular formation. Specifically, the reticular formation at the midbrain level has been found to maintain a facilitatory or augmenting influence on motor reflexes. However, the more caudal portion of the reticular formation in the pons and medulla has been found to have an inhibitory role, keeping motor activity under control [64].

2.6.2 Ascending Reticular Activating System

The reticular formation and the ARAS are both necessary for arousal of the brain for attention, perception and conscious learning. Without the collaboration of the ARAS the oral sensory messages cannot be projected from the reticular formation to the cortex and hence, they cannot be elaborated and decoded, resulting in no discrimination of the senses. Thus, the state of excitation and adjustment or adaptation level of the reticular formation as it monitors both incoming sensory and outgoing motor messages, as well as the adaptation of the ARAS, become part of the process of arousal of the brain, for learning ability and habit formation [63, 64].

These changes of the reticular formation and of the ARAS depend upon the ebb and flow of activity in the afferent and efferent systems. The strategic location of the ARAS at the cross roads of the sensory input and motor output systems of the cranial nerves, including the trigeminal input and output pathways, permits the ARAS to sample and monitor all such activities. In doing so it becomes adjusted to certain levels of activity and its own response is projected to the cerebral cortex to exploit aspects of its information-processing capabilities, such as perception, learning, memory, motor planning, etc. The regulation of mature oral sensorimotor functions derive from the cerebral cortex through the ARAS. The sensory-input/processing function of the cortex determines the level of arousal of the brain for attention and awareness. For example, if the ARAS is activated experimentally by electrical stimulation, the cortex shifts from sleep to waking in the encephalogram. Thus, the ARAS is involved in the sleep and waking rhythm of the organism, which is controlled by the cerebral cortex through the neural activity of the sensory-input/information-processing function [63, 64].

When the ebb and flow of activity in the afferent and efferent systems is restricted, compensatory adjustments are made within limits in the reticular formation and the ARAS. When this fails, for instance in sensory overload or sensory deprivation, the cortical factors are not under the control of the ARAS. When this happens persistently, perception is disrupted, attention gives way to distractability and interest to boredom. Behavior, including oral motor behavior, becomes disorganized. This seems to be reflected in either a pervasive inhibition of sensory stimuli or a marked facilitation of oral reflexes with excessive motor activity of a highly stereotyped type and nonadaptive nature, which in turn may lead to oral habit formation [64].

In this view, early sensory deprivation of the brain or early growth and development in an impoverished environment, one with diminished heterogeneity and a reduced set of opportunities for manipulation and discrimination, not only rob the organism–mouth of the opportunity for constructing sensory models of the environment in order to deal efficiently with them, but it also prevents the utilization of sensory modalities to extract and evaluate useful information from the environment [69, 70]. This may result in impairment of sensorimotor functions through the disruption of cortical space and time quality organization (perceptual changes), as well as deterioration in learning ability in association with a decline in reasoning (cognitive changes) [64]. This disorientation of oral motor behavior is often associated with a reduction in taste, poor adaptive capacity of the occlusion of the teeth and a decline in function of the mouth. In general irritability and restlessness of the mouth are coupled with anxiety [63, 69].

Sensory deprivation may occur experimentally with unvarying sensory stimulation of the cortex by reducing both the amplitude and the rate of change of the stimuli to a monotonous level, at which the sensory receptors adapt and stop firing resulting in sensory deprivation of the brain, which is characterized by disruption of the capacity to learn or even to think [71, 72]. This breakdown of the sensory–perceptual–motor and learning–memory references, by which the organism–mouth guides its correction strategies in perceiving, becoming cognizant, reacting and manipulating the environment [69], and guides the ontogenetic development of the body [73], may make it increasingly difficult in early postnatal development for the oral sensorimotor functions to structure a normal anatomical mouth [43, 65].

Specifically, since normal perception depends partly on normal motor activity [16], and normal motor activity generates normal patterned sensory input [58], resulting in normal development of the brain [33], then it is likely that the early deviation of the mouth from the normal sensory-input perceptual function of the cortex will affect the brain structure and its cognitive state. These changes in turn will be reflected in the neurological development of the mouth. In this view, Bosma [43, 65] suggested that the developing central nervous system guides the development of the anatomical mouth and the position of the teeth. Conversely, the mouth is developmentally inscribed at all stages upon the maturing brain and the neural networks of the brain are modulated by the oral sensory experiences.

2.6.3 Corticoreticular Control

It is established functionally that there are interconnections between the cortex and the reticular formation. This means that arousal and alerting of the brain are not solely dependent upon peripheral sensory influx, but may equally well be produced by cortical activity [64]. In this view, imagination of past or present stimuli has the potential to excite the reticular formation and the ARAS, which in turn produces a suitable state of activity in the cortex. The cortical factors now not under the control of ARAS may play a more prominent role. The effect tends to be facilitation of stimuli, which will favor cortical restless hyperactivity [15, 64]. Recently, it has been shown that the motor cortex is activated as if sensory input had occurred in anticipation of a motor activity [74]. This shows that we experience sensory feedback before the sensory input occurs [15].

2.6.4 Centrifugal Afferent Control and Central Inhibitory Mechanisms

According to most neurologists the centrifugal influences serve the purpose of making the receptors or pathways more capable of sending information or of preventing the normal deluge of sensory impulses from reaching the higher central nervous system levels. In this view, the centrifugal afferent control system can reduce the sensory input if the motor response output becomes excessive. The descending fibers from the cortex terminate in excitatory synapses with interneurons, which in turn act through presynaptic or postsynaptic inhibition on the neurons of the ascending affected pathways. The effect of centrifugal afferent control is generally inhibitory. It is exerted at the synaptic relays, as well as at the receptors themselves, tending to suppress the amount of information transmitted to the cortex [64]. Under centrifugal afferent control there is a marked reduction in the total mass of the sensory input, while the "useful" or "significant" information is preserved or even enhanced. Thus, selectivity or filtering is an important consequence of these inhibitory mechanisms. It is of interest that the centrifugal control mechanisms are learned with early sensory experience [64].

The importance of inhibitory reflex mechanisms in the normal motor maturation, is seen in the early postnatal period. For example, if one gives a two-year-old child a pencil, it holds it tightly. It is part of the educational program of the school to weaken and inhibit the grasp reflex, so that the child will be able to hold the pencil and write with it. Thus the major portion of the education that human beings receive during the preschool years, in school and in later life, consists in the development of inhibitory nerve fibers and inhibition mechanisms in order to overcome reflexes. A new-born child cries when it is hungry. It must get used to eating at meal times. Thus, one of the first things a child learns is to suppress the reflex to cry when hunger pangs appear. Then a child is taught to inhibit the bladder and intestinal reflexes, and to let them function only at certain times. This is the beginning of training for life, and this is how it continues for ten, twenty, thirty years under the constant motto "develop inhibitory fibers," so that training may be said to be almost equivalent to the possession of inhibitory mechanisms, and the ability to control oral and body reflexes, drives and animal instincts in social behavior by means of them [37, 75].

Childhood is generally characterized by extremely extensive reflex movements arising in response to relatively weak (from the adult point of view) external sensory stimulation. Gradually, however, during development one or two groups of muscles separate from the mass of other muscles and the movements become more limited. Having become more limited, the movements acquire a definite character. It is in this limiting process that the inhibitory mechanisms take part [37]. For example, the normal coordination of the contraction of the muscle groups involved in mastication and deglutition require central inhibition of the neurons of the trigeminal nuclei when the deglutition function is initiated, and conversely central inhibition of the glossopharyngeal motor neurons, which innervate the pharyngeal muscles of deglutition, when the masticatory muscle group is activated [76].

Thus, by exerting presynaptic and postsynaptic inhibition, the cerebral cortex is able to block the synapses in the sensory pathways and hence, give the opportunity for an inhibitory action to occur, that sharpens the neural signals by eliminating the weaker and irrelevant excitatory stimuli. As a consequence, the stimuli can be more precisely located and evaluated by the brain [37].

In fact, a tactile stimulus, for instance, not only produces a sensation but at the same time produces an inhibition of the surrounding cutaneous or mucosal area [44]. Thus, the concept maintained in telephone engineering—that we have a simple input and output—does not hold in the sensory systems, since every stimulus or input produces definitely two phenomena. There is a usual output as found in telephone systems, but there is also an inhibition in the surrounding area of the stimulus produced by the output [77].

In this view, the brain makes a simplification by blocking certain excitatory inputs in order to protect the masticatory apparatus from the many reflexes that would normally arise from all structures had been stimulated separately during mastication. This complex interaction of excitatory and inhibitory influences upon the oral tissues can limit the forces developed during chewing. For example, when crushing of a food bolus is desired, a protective mechanism involving a cortical loop may be called into play and periodontally induce inhibition of jaw-closing motoneurons, which outweighs the excitatory influence, resulting in cessation of jaw-closing muscle activity during a masticatory cycle [78].

It is also interesting to note that when we are very intensely occupied, for instance in carrying out some action or in experiencing or in thinking out some action, the cerebral cortex can block the synapses in order to protect itself from being bothered by stimuli that can be neglected. Accordingly, in the heat of combat, severe injuries can be ignored by the individual by suppression of the pain pathway to the brain. Thus, we can account of afferent anesthesia, of hypnosis or yoga or acupuncture by the cerebral cortex inhibiting the sensory pathways [37].

3 Sensory Deprivation of the Brain

3.1 Introduction

The term sensory deprivation carries the implication that the tissues of the central nervous system can experience a peculiar kind of need or appetite arising out of deprivation of sensory stimulation of the brain, which is comparable to the need of dehydrated tissues for water. The term implies that the brain must have a certain minimal patterned continuous inflow of sensory experience, and that whenever the sensory input drops below a critical level for some time, a hunger arises in the brain for more stimulation, in order to influence and/or maintain the arousal reaction of the brain for attention, awareness, and learning, since the deafferented brain is the sleeping brain [71].

Some simple experimental work based on this assumption was done by Levy [80] on oral stimulation in chickens and humans. The implication was that if the oral stimulation dropped below a certain level, the infant or chick developed a persistent hangover of excessive need for more oral stimulation.

The existing evidence now suggests that sensory systems, like the mouth, being particularly sensitive to stimuli discharge more frequently when there is variation in the sensory stimuli, rather than when there is not [71, 81, 82]. This is because the frequency of action potentials in the sensory nerves diminishes during continuous stimulation by the same stimulus of the same intensity (monotonous stimulation), leading to adaptation of the sensory receptors to the stimulus which stop firing [44]. Thus, the capacity of a stimulus to evoke and maintain arousal of the brain for attention, awareness and learning is lost when the same stimulus of the same intensity is repeatedly (monotonously) presented for some time [71, 81].

Kubie [71] experimentally produced adaptation of sensory receptors (touch, temperature, pressure, proprioceptor) by reducing the sensory flow to a monotonous level, that is by reducing the amplitude and the rate of change of stimulus to levels at which adaptation of receptors occurred and they stopped firing their nerves. He then suggested that the effectiveness of a stimulus is not related to the volume of the input or the direction of change, but to its amplitude and rate of change.

Freedman et al. [82] further expanded the term sensory deprivation to cover two different situations: (1) that in which an attempt is made to reduce the absolute level of sensory

M. Z. Pimenidis, *The Neurobiology of Orthodontics*,
DOI: 10.1007/978-3-642-00396-7_3, © Springer Verlag Berlin Heidelberg 2009

input to a minimum, called "deprivation," and (2) that in which an attempt is made to eliminate order and meaning from the sensory input, called "not patterned" input.

The process of incorporating ordered relationships into a personal perceptual scheme is one of extracting useful information from the welter of "noisy" sensations with which the organism or the mouth is constantly bombarded. The world of the infant is a buzzing confusion, because he has not yet learned which of the incoming stimuli are meaningful and which are not. If the infant's oral sensory cues are to acquire meaning and be incorporated into the scheme of the mouth, they must be regular, not changing capriciously or adventitiously [82].

In addition, Freedman et al. [82] suggested that the perception of environmental relationships is intermodal as well as intramodal, implying that perceptual schemata for different sense modalities are closely tied together and may themselves be used to reaffirm one another. For example, we see a cup of coffee, then reach out our hand to pick it up. Success in reaching and touching the cup provides kinesthetic and tactile confirmation of the visually "measured" position and distance of the cup.

A number of investigators have speculated about the neurological effects produced by sensory deprivation. Heron [81] suggested that sensory deprivation seems to involve a general disorganization of brain function, similar to that produced by anoxia, by large brain tumors, or by administration of certain drugs.

Hebb [16] considered the effects of monotonous sensory stimulation on learned behavior and suggested that this condition produces disruption of the capacity of the brain to learn or even to think. He pointed out the need for patterned sensory stimulation of the brain, and he stressed that in the absence of varied stimulation the function of the brain becomes less efficient, resulting in a wide variety of behavioral, motor, perceptual, and cognitive changes as well as habit formation produced on the basis of the reduced sensory input (sensory deprivation of brain), through the dysfunction of the ascending reticular activating system (ARAS). He concluded by stating that only patterned (ordered, meaningful) sensory stimuli of varying intensity seem to maintain the strength of the synapses, thus enhancing the processing of information, which contributes to the richness of sensory experience.

This means that the same synapse may undergo graded weakening or strengthening according to the type of stimulus (monotonous stimulus vs. varied and ordered stimulus), which in turn determines the release of chemical signal molecules (neurotransmitters, growth factors) strengthening synapses and actively involved in the developing of neural input circuits [2].

The neural activities which arise in the brain through the strengthening of synapses and of the patterned sensory input in turn underlie sensation, perception, learning, and memory of experiences, as well as motor control of muscles, and it is these mental activities that define the nature of brain changes, through the normal stimulation of the senses, including those of the mouth [15, 58].

3.2 Sensory Deprivation and Thumb-Sucking

Thumb- and finger-sucking habits appear to be relatively prevalent in infancy and childhood, and there appears to be some relationship between the habits and the development or exacerbation of malocclusions. In general, however, there is agreement that these habits are unlikely to result in permanent damage to the dentition, particularly if they are abandoned by the age of four to five years. If the child persists beyond this period in thumb- or finger-sucking, it is felt that the likelihood of harmful effects is increased, although the extent to which they are increased is not fully known [83].

Thus, an alternative consideration is required if the effect of thumb-sucking on the occlusion of the teeth or on oral motor behavior is to be understood. A number of writers have suggested that the heredity component of malocclusion seems to be more important for the development of later malocclusions than does the habit of persistent sucking itself [83, 84].

The potential influence, however, of oral habits, for instance thumb-sucking, on the etiology of malocclusions of the teeth in early childhood, might have to be reconsidered from the point of view that the habit may impose conditions of sensory deprivation of the cerebral cortex, affecting the normal sensorimotor integration function, which in turn may result in abnormal neuromuscular function of the mouth.

Specifically, a number of researchers have provided evidence that the sensory systems, including the mouth, being particularly sensitive to stimuli discharge more frequently when there is variation in the sensory stimuli, rather than when there is not [71, 81, 82]. This is because the frequency of action potentials in the sensory nerves that innervate the receptors diminishes during continuous stimulation by the same stimulus of the same intensity (monotonous stimulation) leading to adaptation (fatigue) of the sensory receptors of the mouth, which stop firing [44]. Thus, very little sensory information is transmitted to the brain, resulting in sensory deprivation of the somesthetic cortex [71, 82]. These cortical changes may in turn prevent the massive synchronous depolarization of the membrane of cortical neurons (neurons do not fire in synchrony), which is a prerequisite for the generation of the conscious electromagnetic field, the physical substrate of conscious motor control of muscles, resulting in motor dysfunction of the trigeminal system [85] (see Sect. 4.8).

In this view, the repetitive monotonous stimulation of the oral senses through thumb-sucking (the same stimulus of the same intensity is continually repeated for a long time) may induce adaptation (fatigue) and decline of the sensibility of receptors of the mouth, which eventually stop firing, depriving the cortex of the normal sensory input, resulting in impairment of learning ability, perception, and motor control of muscles [16, 82].

The fact that a strong correlation exists between thumb-sucking and the class II division 1 malocclusion of the teeth, in which the mandible is abnormally held by the muscles in a retruded occlusal position [1], may be related to oral disability of learning motor functions as a consequence of sensory deprivation, induced by the thumb-sucking.

Sensory deprivation of the brain results through the affected function of the ARAS of the reticular formation. The normal function of the ARAS is necessary for arousal of the brain for attention, awareness, and conscious learning. Without the collaboration of the ARAS the oral sensory information cannot be projected from the reticular formation to the cortex. This results in no discrimination of the oral senses and hence, sensory deprivation. Normal regulation of oral sensorimotor functions derive from the cerebral cortex through the ARAS. When the ARAS is affected by the monotonous stimulation of the oral senses through the habit activity, perception is disrupted and attention gives way to distractability. Behavior, including oral behavior becomes disorganized. This seems to be reflected in either a pervasive inhibition of sensory stimuli or a marked facilitation of oral reflexes with excessive motor activity of highly stereotyped type and of no adaptive nature, which in turn, may lead to habit formation [64]. Thus thumb-sucking may induce sensory deprivation and conversely oral sensory deprivation may induce the habit activity through the affected ARAS.

The fact that many children suck their fingers in order to reduce mental stress in order to sleep may suggest that these children seek isolation from the environment, and that the thumb-sucking activity helps towards the reduction of the input of external stimuli to their brain. In other words these children are experiencing a "functional deafferentation" of the cortex, which is the substrate of the sleeping brain [71, 86].

Thus, the habit of thumb-sucking may emphasize what Kubie [71] and Heron [81] have suggested, that the capacity of a stimulus to evoke and maintain arousal of the brain for attention, awareness, learning, perception, and motor control of muscles, is lost when the stimulus is monotonously (not patterned input – without order and meaning for the cortex) presented for some time. With the breakdown of the perception (internal frame of reference of the external world) it becomes increasing difficult for the brain to structure the normal physiology and anatomy of the mouth [82].

3.3 Sensory Deprivation and Malocclusions of the Teeth

The concurrently developing brain and mouth are liable to several patterns of abnormality and impairment. Specifically, understanding of oral sensation and perception has been substantially advanced by studies of anomalous and neurologically impaired subjects [43, 65]. In the following paragraphs a disruption of the central representation of the teeth and of oral motor functions is described in a case of malocclusion of the teeth.

Specifically, an abnormal perception of the incisor teeth associated with abnormal oral motor behavior, such as tongue thrust and poorly coordinated masticatory movements, was diagnosed in a 21-year-old male patient with severe class III anterior open bite malocclusion of the teeth. The patient was not aware of the presence of the incisor teeth in his mouth without the use of vision, i.e., he had the feeling that the incisor teeth were missing [87]. However, after the orthodontic closure of the open bite and the establishment of normal occlusion of teeth, the patient become conscious of the presence of the full complement of teeth in his mouth (Figs. 3.1–3.3).

Accordingly, it was postulated that the impairment of perception of the teeth may indicate that in the open bite the sensory input, stemming at least from the periodontal mechanoreceptors of the incisor teeth, was reduced resulting in sensory deprivation of the somesthetic cortex, where the oral sensory system is point-to-point represented and perceived. This view, is consistent with the studies of Morimoto and Kawamura [88] who found that subjects with a normal occlusal relationship of their incisor teeth could discriminate a 0.2-mm difference in the size of wires held between the teeth, while subjects with maloccluded incisor teeth needed a difference of more than 0.4-mm in wire size in order to discriminate them. They concluded that periodontal receptors have reduced sensibility in malopposed teeth.

Similarly, the studies of Teuber et al. [89] and Haber [90] suggest that changes in perceptual function may happen in limited areas of the somesthetic cortex whose sensory

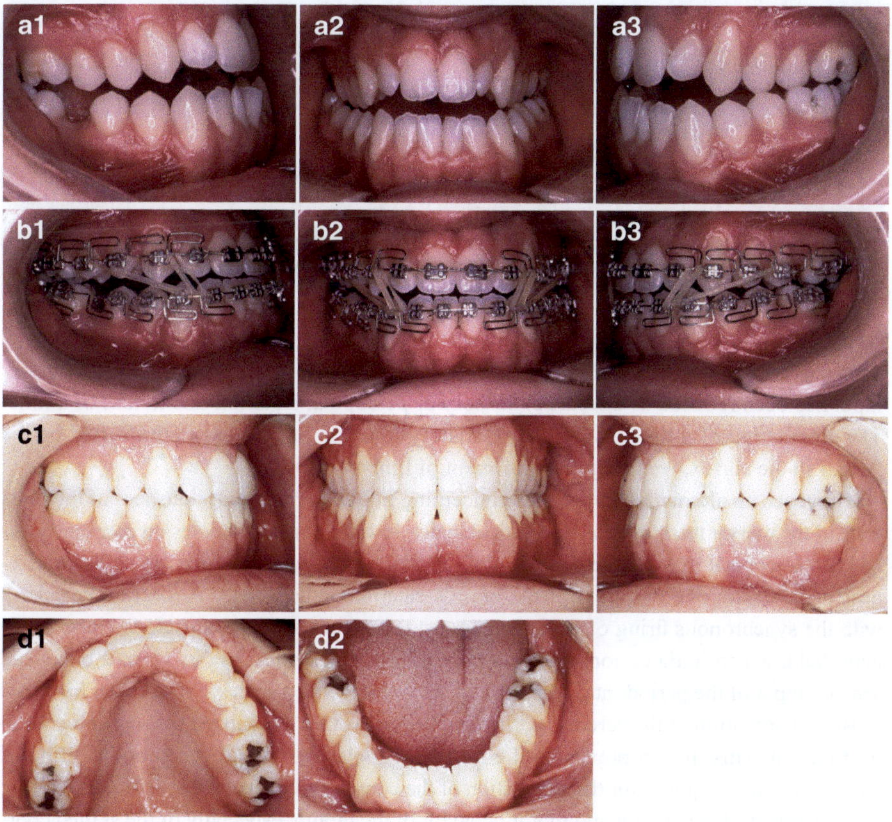

Fig. 3.1 Intraoral photographs. **(a1–a3)** Pretreatment. **(b1–b3)** After a year of active treatment, upper and lower MEAW (multiloop edge-wise arch wires) were placed and 3/16-inch, 6-oz vertical and class III elastics were applied. **(c1–c3)**, **(d1, d2,)** Posttreatment. The active treatment duration was 32 months

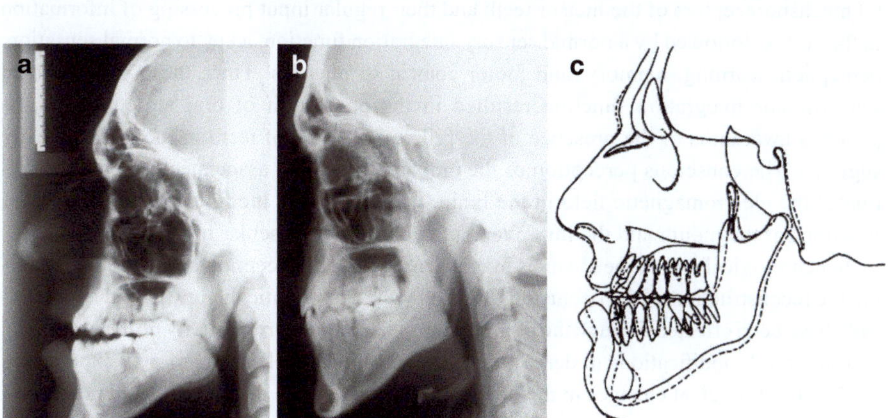

Fig. 3.2 Cephalometric radiographs. **(a)** Pretreatment. **(b)** Posttreatment. **(c)** Pre-/posttreatment superimposition

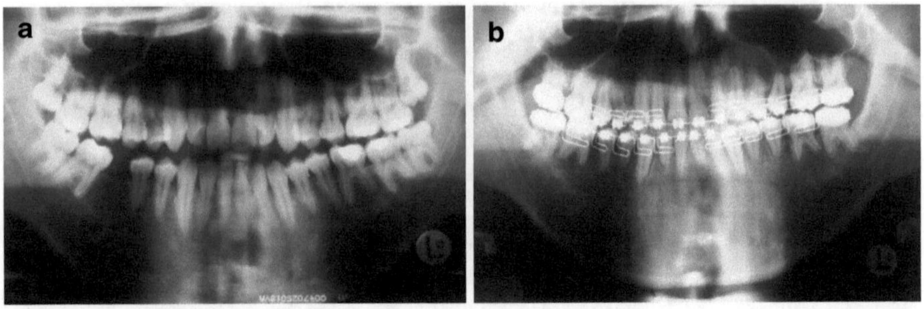

Fig. 3.3 Pantomographs. **(a)** Pretreatment. **(b)** Just before appliance removal

input has been reduced as a result of sensory or perceptual isolation due to loss of a limb. The amputation of a limb affects the regulatory function of sensory input exerted by the ARAS in the cortex, resulting in impairment of limb perception.

However, an alternative hypothesis concerning impairment of consciousness of the presence of the incisor teeth in open bite malocclusion based on current studies on consciousness and motor control may be the following. McFadden [85] suggested that conscious awareness or perception is a product of the brain's electromagnetic field. Perception correlates with the synchronous firing of cortical neurons, thus generating the conscious electromagnetic field, which initiates conscious learning of motor activities. In this view, the reduced sensory input of the periodontal receptors of the incisor teeth may have resulted in localized sensory deprivation of the cerebral cortex and hence in at least localized sensory integration dysfunction. This in turn prevented the massive depolarization and synchronous firing of cortical neurons required for the generation of the conscious electromagnetic field, resulting in impairment of perception and motor behavior of the mouth, especially of the skilled coordinated movements of chewing.

However, the closure of the open bite and the establishment of normal occlusion of the teeth, through the orthodontic treatment, resulted in the normal sensibility of the periodontal mechanoreceptors of the incisor teeth and their regular input processing of information in the cortex, followed by a normal sensory integration function, a key to normal sensation, perception, learning, memory, and motor control of muscles. Thus, the normal cortical sensorimotor integration function resulted in the perception of oral structures, as the patient's awareness of the presence of the full complement of teeth in the mouth clearly suggests. The conscious perception of the oral structures was associated with the generation of the electromagnetic field in the brain, which initiated the relearning of the motor functions of the mouth and the improvement in oral neuromuscular function.

In neurological terms, the changes in oral sensation and perception of the mouth following the reconstruction of the occlusion of teeth by the orthodontic therapy, could hypothetically have been effected through the brain's neuroplasticity. Neuroplasticity implies chemical and structural modifications in dendritic spines in response to learning of new experiences [3–5]. Morphological changes in the dendritic spine are key to its plasticity and processing [6, 7]. Conversely, electrical stimulation leading to long-term potentiation (LTP), leads to synaptogenesis, i.e., alteration in the number of spines by the formation of new ones [8, 41].

Spines on dendrites are considered as the major sites of presynaptic excitatory inputs of the brain [3, 15].

Thus, the orthodontic procedures might have stimulated the oral mechanosensory systems so that afferent nerves carrying action potentials from periodontal periosteum, muscles, tendons, oral mucosa, receptors, etc, converged to the dendritic spines of the cortical neurons, where they were pooled and integrated in the microtubules of the subsynaptic zone lying beneath the spinous synapses. The subsynaptic zone of dendrites gives rise to spines and it is the physical substrate of memory storage of new learned experiences [15, 38]. The resulting new frequencies of muscle information contraction patterns changed the oral motor behavior [85]. Bone reorganization in turn, was "tuned" to the new frequencies of muscles through bone mechanosensation and intraosseous production of electric fields traveling in bone cells connected by gap junctions (electrical synapses) [91].

In addition, the awareness of the presence of the full complement of teeth in the patient's mouth suggests that the perceptual scheme of the teeth has been inscribed in memory. Bliss and Lomo [17] discovered that in the hippocampus memory center there are special neuronal circuits serving long-term memory (LTP), and that brief experiences can induce memory storage lasting for a long time after the stimulus has been stopped.

In sum, the "orthodontic stimulus" has induced learning of new sensory-perceptual and motor functions of the mouth through the transformations of the corresponding sensory-perceptual, motor and memory areas of the brain. This might imply that the orthodontic stimulus has activated specific brain genes expressing plasticity. This might in turn imply that the rules that govern the structure and function of sensory and motor maps in the cortex can be decoded by the orthodontic stimulus, suggesting that the formation of cortical maps is experience-driven.

This insight is emancipating, because most molecular biologists were certain that environmental factors, such as tactile and pressure experiences could not affect gene function [69, 73]. Schanberg [73], however, located a growth gene that is responsible for the relationship between touch stimulation and growth. Schanberg found that touch deprivation of rat pups by their mothers resulted in growth deprivation through endocrine hormone effects. He then suggested that the brain can regulate growth genes through sensory (touch) processes affecting the production of growth hormone. Similarly, Moss [91] suggested that mechanotransduction of stimuli and computational bone biology offers an explanatory chain extending from the epigenetic event of muscle contraction hierarchically downward to the regulation of the bone cell genome.

3.4 Sensory Deprivation and Occlusal Interferences

The presence of irregularities of the teeth with premature contacts during closure, has long been considered to be a key factor in oral neuromuscular dysfunction, irrespective of whether it is assumed to operate by central initiation of reflex hyperactivity or by causing local hyperactivity of muscles at the brainstem level [95]. However, although many patients with this dysfunction do possess a defective occlusion, so too do a similar proportion of the population without this dysfunction [96]. This means that there is not a direct cause and effect relationship between occlusion pattern and dysfunction [95].

In the following paragraphs an attempt is made to relate occlusal interferences or premature dental contacts and the associated neuromuscular dysfunction, to sensory deprivation of the cerebral cortex. Sensory deprivation causes breakdown of the relative balance between the function of the ARAS and of cortical activity. Consequently, the individual becomes more alert, and this stage of the brain tends to facilitate the irrelevant stimuli, resulting in hyperactivity of muscles [64].

Recent neurophysiological studies of forces applied to the crown of teeth (tooth loads are essential for the fine motor control of the jaw muscles during mastication) have indicated that the periodontal mechanoreceptors are directionally sensitive to tooth loading. For example, the receptors of the incisor teeth and premolars are most sensitive (discharge vigorously) to forces applied horizontally to their crowns. This means that forces of the same magnitude applied in other directions do not evoke as great a response as the horizontal forces. This implies that fewer receptors respond in non-horizontal forces, providing ambiguous information to the central nervous system about the amplitude of the force [97, 98].

In addition, other neurophysiological studies have indicated that the periodontal mechanoreceptors signal only relatively small forces and saturate when larger forces are applied to the teeth, resulting in no useful information to the brain as to their magnitude. Thus, the sustained force level below which the mechanoreceptors signal is about 1 N (corresponding to 98 g weight) for the anterior teeth and about 3–4 N for the posterior teeth [99].

Accordingly, when a force of, for instance, 1 N is applied to an anterior tooth, the tooth moves slightly in the socket. Clinically, this sometimes is observed as fremitus which induces stress and strain in the periodontal ligament. A depolarizing receptor potential and an action potential are generated in the receptors and in the nerve terminals, respectively. The output then from an individual periodontal mechanoreceptor depends on the effectiveness of the fremitus in producing strain in the periodontal receptors, as well as of the directional sensitivity of the receptors to generate a vigorous action potential in the nerve terminals [93].

Thus a large number of mechanical factors determine the effectiveness of the movement of a tooth producing strain in the periodontal ligament and stimulation of receptors. These factors include the size of the force, the size and shape of the tooth, the point of rotation (fulcrum) of the tooth, the point of application of the force, and the presence of adjacent teeth with proximal contacts. The effect of these factors on the receptor depends on the precise location of the receptors about the root, the receptors' orientation with respect to the surrounding collagen fibers under tension, and the viscoelastic properties of the periodontal ligament [93]. As a consequence of these factors each receptor is optimally stimulated by forces applied in directions that most effectively strain the receptors (directional sensitivity) [93].

It is also noted that in the normal dentition the receptive field of the periodontal mechanoreceptors extends beyond a single tooth to adjacent teeth through the interdental contact and the transeptal collagen fibers, resulting in a multitooth receptive field response [100]. The receptive field, however, might be even greater considering the fact that the bone cells (except the osteoclasts) are connected by gap junctions thus forming large cellular networks (syncytium), so that the electric fields generated by the physical loading of the alveolar bone are "tunneling" through the gap junctions into the entire bone cellular network [91]. In malocclusions of the teeth, however, only the "high spots" are in contact during closure. The proximal contacts are lost, because the teeth involved in the premature contacts are hypererupted. Consequently, the multitooth receptive field response might be considerably reduced, resulting in very little sensory input to the brain.

Also, there might be an additional reason contributing to the reduction in the sensory input, as follows. The force exerted on a hypererupted tooth during chewing or closure of the teeth might be higher than the level of loading the receptors can tolerate. For instance, if the force that is applied on a high spot in the anterior segment of the dentition is higher than 1 N, then its periodontal receptors saturate and provide no useful information to the brain [99]. Accordingly, the cerebral cortex may experience localized sensory deprivation, considering the large innervation and receptive field of the anterior teeth, resulting in neuromuscular dysfunction.

It is interesting that the hypererupted teeth act as dangerous high spots in the sensorimotor system of the mouth. The ARAS is implicated in alerting the brain to the potential danger [64]. This means that the central nervous system geared to fight or flight, may activate the protective system, while at the same time inhibiting the discriminative system on which the integration function of the cortex is based. In the protective system the tactile stimuli of the mouth are interpreted as nociceptive stimuli. The tactile system is prepared to cope with the fight or flight situation. The other senses, which form the discriminative system play now a lesser role and their input is inhibited [101]. This may result in sensory deprivation of the cortex and sensorimotor dysfunction of the mouth (see Sect. 8.30).

The fear of biting on a high spot produces adrenaline, which enhances readiness for fight or flight, but also lowers the threshold of tactile receptors (increase in sensitivity), thus enhancing their role in the protective system against the discriminative system. On the other hand, the emotional involvement complicates further the function of the receptors, as discussed below, leading to sensory deprivation.

However, there is another reason why the premature dental contacts might cause sensory deprivation. In this view, the unvarying frequency of the stimulus has to be considered. The same stimulus of the same intensity is continually (monotonously) repeated on the teeth with premature contact leading to adaptation (fatigue) of periodontal mechanoreceptors. The receptors stop firing and the cerebral cortex suffers sensory deprivation (see Sect. 3.7). The ARAS can adjust its function to the reduction in sensory input within limits [64] (see Sect. 2.6).

However, if into the already sensory-deprived brain because of the premature dental contacts, is introduced an additional derangement of the ARAS and of cortical activity, for instance in the form of the habit of thumb-sucking (thumb-sucking is very common in young children, and is strongly correlated with class II division 1 malocclusion of teeth [1]), then sensory overload may occur in the brain. Sensory overload is an abnormal condition that develops through intense multisensory experiences (touch, pressure, proprioception, etc) coming from the same source, in this instance, from the mouth. These stimuli due to their unvarying frequency, intensity and complexity are unwelcome to the brain, resulting in sensory deprivation, followed by deficits in sensation, perception, learning, memory, and motor control [63, 64].

Accordingly, since perception, learning, and memory are correlated neural activities [15, 69] and related to emotional development [63], it is not unlikely that the degree of perceptual deficit in sensory overload will also have an emotional effect in children, for instance anxiety, which very often is associated with thumb-sucking. Anxiety or mental stress may in turn aggravate the sensory deprivation of the brain, as discussed below. Finally, in people with malocclusions of the teeth the sensibility of the teeth is reduced and they have a poor ability to discriminate stimuli between the teeth. This means that normal occlusion of the teeth is necessary to maintain the complete normal periodontal sensibility [88].

In sum, the occlusal interferences or premature contacts of malocclusions of the teeth, as factors causing neuromuscular dysfunction, may have to be reinterpreted through the sensory deprivation mechanism, which affects the normal sensory input and thus the sensorimotor integration function of the cerebral cortex. The reduced oral sensory input affects the regulatory function of the ARAS upon the cortex, especially when the central derangement is alerting the protective mechanisms of the brain, while the discriminative system is suppressed. The central insults prevent any adaptational effort of the oral sensorimotor system, resulting in complete disorganization of oral motor behavior.

When sensory deprivation occurs in the period of early postnatal growth and development, the learning of skilled oral movements, such as chewing, by individual central design may not occur, resulting in uncoordinated muscular patterns [102]. Instead the primary, unlearned, stereotyped patterns of movements may still be present in young children, suggesting cortical cognitive and motor dysfunction [102].

In this context, early malocclusions of the teeth associated with sensory deprivation may lead to many forms of neuromuscular dysfunction and abnormal bone growth. For example, Moss [91] suggested that mechanosensory processes operate in bone, which is electrically active. Bone is "tuned" to the precise frequencies of skeletal muscle activity, and has morphogenetic influences. In this view, central conscious electromagnetic information output to muscles may regulate the contraction frequency of muscles [85], which in turn may influence bone organization. Thus, normal or abnormal cortical pyramidal motor neuron output may influence normal or abnormal neuromuscular patterns and bone growth. Normal neuromuscular patterns require an intact oral afferent nerve input function, which is a significant morphogenetic factor in the development of the oral region [43].

3.5 Sensory Deprivation and Emotions in the Deciduous and Mixed Dentition

The reduction in sensory input in the cerebral cortex, through decreased or unvarying frequency sensory input, may lead to changes in emotion, and emotional changes may lead to alteration in the function of sensory receptors and perception, through hypothalamic and limbic system mechanisms. In this view, the sensory, parasympathetic and sympathetic branches of the nervous system are related [63, 64].

The interaction of the sensory deprivation of the cortex with the autonomic nervous system may be relevant to the emotions of children during their preschool education. In this period we are learning most of our experiences and hence, sensory deprivation of the brain is critical to learning ability [15]. Education at this age, however, means limitation and therefore it is a period of emotional loading and psychic narrowing [79]. The children may suffer from mental distress, which may lead to changes in receptor function, followed by sensory deprivation of the brain and oral sensorimotor dysfunction [63]. The impairment of the mobility of the mouth may in turn restrict further the stimulation of oral receptors and hence aggravate thc sensory deprivation, in accordance with the principle that normal oral motions generate most of the sensory input [43].

In addition, most mentally stressed children may develop severe nocturnal bruxism and/or thumb-sucking [83]. Thumb-sucking may lead to sensory deprivation, as already discussed. Bruxism is viewed as a central disorder involving excessive activation of the "central pattern generator" or "chewing center" [78].

Lindsley [64] suggested that when the arousal of the cerebral cortex is diminished through the mechanisms affected by the sensory deprivation, the ARAS, the learning and perceptual abilities of the cortex are impaired. This breakdown of the relative balance between the function of the ARAS and cortical activity is believed to be associated with some of the motor behavior changes seen in sensory deprivation, including the formation of habits. In other words the individual becomes more alert, which is characterized by hyperactivity, because the cortical factors are not under the regulatory influence of the ARAS. Thus, Lindsley provides a way of understanding the hyperactivity of muscles in bruxism and in thumb-sucking through the sensory deprivation of the cortex.

On the other hand, it has been suggested that the abrasion of deciduous teeth during bruxism helps the mandible shift from a retrognathic position during the period of the deciduous dentition to an anterior normal developmental position when the mixed dentition is present [94]. It can be argued, however, that during the abrasion period of teeth the child is actively involved in bruxism and/or thumb-sucking, which are associated with emotional loading. These nervous system derangements, however, have been implicated, as already discussed, in the sensory deprivation of the cortex and sensory integration dysfunction, which are followed by abnormal oral neuromuscular function. Thus, impairment of normal function of the muscles may prevent the protrusive movement of the mandible during the subsequent growth and development period of the mouth and instead may lead to abnormal retrognathic growth of the mandible, resulting in the familiar class II division 1 malocclusion of the teeth. A similar assumption was made by Moyers [1] who suggested that the emotional status of young children may reflexly guide the mandible in an abnormal occlusal position, which in turn, may lead to an abnormal growth pattern.

Children may also experience emotional disturbances and sensory deprivation during eruption of the first permanent molar teeth with their distinct cusp system. Children then experience occlusal interferences and the chewing and speech functions have to adapt to these rigid structures limiting the freedom of mandibular movements [94]. At the same time children are exposed to an increased social narrowing. They are going to school and have to change from a life of standing, jumping, running to a life of sitting. Thus, children may again suffer mental distress which may lead to changes in receptors and sensory deprivation which in turn may affect the function of the deep limbic system as discussed below.

3.6 Looking into Love and Depression: The Deep Limbic System

Every time children have an angry thought, an unkind thought, a sad thought, or a cranky thought, their brain releases chemicals that make their body feel bad and activate the deep limbic system. When children are angry their muscles become tense, their hearts beat faster, their hands start to sweat, and they may even begin to feel a little dizzy. Their body reacts to every negative thought they have. A negative thought is like pollution to the brain and body. When children are happy, their muscles relax, their hearts beat more slowly, they breath more slowly and their body reacts to good thoughts. Of significant importance is the bonding between children and parents. Children need actual physical time with their parents (limbic bonding). Resnick et al. [150] have reported that teenagers who felt loved and connected to their parents have a significantly lower incidence of pregnancy, drug use, violence and suicide.

The classic term limbic system incorporates the cingulate gyrus and deep temporal lobes. The term "deep limbic system" includes the thalamic structures and hypothalamus, along with the surrounding structures [199]. Emotional disturbances of children may involve abnormality of the function of the deep limbic system, especially of the hypothalamus. The deep limbic system affects motivation, people sometimes develop an "I don't care" attitude about life and work; they don't have the energy to care. Children, because they feel hopeless about the outcome, may have little will-power to follow through with tasks. Since the sleep and appetite centers are in the deep limbic system, disruption of function can lead to changes, which may mean an inclination toward too much or two little of either. For example, in typical depressive episodes children may lose their appetite and have trouble sleeping despite being chronically tired, and yet in atypical depression they will sleep and eat excessively. In addition, bonding and limbic system problems go hand in hand. For example, the most fundamental bond in the human universe is the mother–infant bond. Hormonal changes after childbirth can cause limbic or emotional problems in the mother. They are called the "baby blues" when they are mild, and postpartum depression or psychosis when they are severe. The mother may emotionally withdraw from the baby, preventing the baby from developing normally [199].

Lack of bonding and depression is known to be caused by a deficit of certain neurotransmitters, especially norepinephrine (noradrenaline) and serotonin. This deficit can cause increased metabolism or inflammation in the deep limbic system, which in turn causes many of the problems associated with depression [199].

3.7 Sensory Adaptation

One of the most important generalizations that can be made about the physiology of the sense organs is adaptation. In effect adaptation in the context of sensation refers to the fact that a prolonged and uniform sensory stimulus eventually ceases to initiate a nerve impulse (action potential); in other words, a sense organ ceases to respond to a prolonged stimulus of uniform intensity. Adaptation is a phenomenon that reminds us that sense organs can respond only to environmental change of state. Thus someone who goes into a room containing a bowl of roses may smell them at first, but then become aware of them. Once this process of sensory adaptation has occurred, no effect of attention can call the smell to mind. It is a popular fallacy that chewing gum regains its flavor if removed from the mouth. What is regained is not the flavor but the ability to taste the flavor as sensory adaptation wears off [142].

A sense organ is especially adapted to respond to one kind of environmental change of state rather than another. For example, to temperature, to light, to pain, sound, posture (as disclosed by muscle and joint receptors), to orientation in space and angular acceleration (both the work of the ear, an equilibratory as well as an acoustic organ). Furthermore, according to the Johannes Müller's Law of Specific Irritability, or Law of Specific Nervous Energy, the modality of sensation whether of light, sound, pain, or posture is determined not by the sense organs, which after all can only generate nerve impulses, but by the nerve itself, or rather by its central connections and integration. Stimulation of the auditory nerve produces sensations of sound, and of the optic nerve sensations of light, by whatever means the nerves are stimulated. Much else in sensation is centrally governed. Thus different cortical receptors provide the perception of a line that is vertical and one that is tilted [142].

Looking into the "Black Box" 4

4.1 Introduction

It has been traditional to view oral motor behavior in terms of a "black box" a term adapted from the physical sciences to designate areas of a system in which it is not known exactly what is taking place. This view describes input and output with little attention paid to what goes on inside the brain. The black box approach of oral motor behavior handles internal processes by simply ignoring them. The brain that regulates the oral sensorimotor behavior, such as speech, chewing, swallowing, secretion of salivary glands, etc, is traditionally viewed as an impenetrable black box receiving inputs (stimuli) and emitting outputs (responses) with little attention paid to what goes inside the brain. Without information about the intervening central mechanisms most early oral motor behavior research contributed little toward the understanding of normal and dysfunctional motor behavior.

I contend that oral motor behaviors (chewing, speech) are different from the motor behaviors performed in other body segments (walking, writing) in daily life, and orthodontics has much to gain by examining the current state of the art in the motor control and learning area. In this chapter an attempt is made to explore the most recent neurophysiological research on the brain's mechanisms that are keys to sensation, perception, learning, memory, and motor behavior functions.

4.2 Decoding the Brain Changes

A growing number of researchers are now focusing their attention on the biophysics of the cytoskeleton of neurons, in order to understand its role in the neurophysiology of sensation and perception. Among the reasons to look beyond the classical models of neurophysiology of perception, is the fact that single-celled organisms, such as the paramecium caudatum, have no neurons or synapses, but still exhibit an apparent awareness of and responsiveness to their environment. Thus, the rudiments of sensation and perception lie someplace other than the complex interactions between neurons and their synapses, although the latter are

M. Z. Pimenidis, *The Neurobiology of Orthodontics*,
DOI: 10.1007/978-3-642-00396-7_4, © Springer Verlag Berlin Heidelberg 2009

certain to contribute to heightening of sensory input and processing of information, through their strengthening [33]. Hebb [16] suggested that the brain strengthens the neuron synapses according to how often they are used (stimulated) through experiences, resulting in heightening of sensory input and processing in the cortex, which in turn underlies sensation, perception, learning, memory, and motor control.

The currently most accepted view is that the processing of information from the senses occurs at the level of the microtubules within the internal cytoskeletal filamentous proteins of the neuron, that include microtubules, actin, and intermediate filaments. In many respects microtubules are considered the control center, which determines the shape, structure, growth, and function of the neuron [15, 33, 38].

The microtubules are cylindrical polymers composed of tubulin dimers (protein building blocks) with plus and minus electrical ends (dipole). Both tubulin ends undergo polymerization–depolymerization cycles during information processing. Tubulin protein is found mainly in the cytoplasm of eukaryotic cells and it is especially enriched in the brain tissue [15].

Microtubules are most stable and have been hypothesized to be central in cellular information processing. Whenever cellular organization and intelligence are required, microtubules are present and involved. For example, there is accumulated evidence that microtubules are computationally relevant to neurocognition [56, 103, 104]. Woolf [15] demonstrated that microtubules and microtubule-associated protein-2 (MAP2) are proteolyzed with learning, as exemplified in hippocampal neurons of rats with contextual fear conditioning. Similarly, Cronly-Dillon and Perry [105] have shown that neurons of the visual cortex produce massive amounts of tubulin during the critical period (from the day the eyes open to postnatal day 35). The critical period is the time during which synaptogenesis and visual learning occurs at highest rates. Thus, tubulin protein is implicated in developmental cognitive processes.

Specifically, the Penrose and Hameroff [113] model of information processing suggests that information storage, representing structural and chemical modifications of microtubules, occurs in the subsynaptic zone lying beneath the spinous synapses of dendrites, wherein microtubules determine synaptic properties of spines based on previous synaptic activity stored as memory. These changes form the physical basis of long-term memory storage in the microtubules of the subsynaptic zone. The morphological changes in dendrite spines, as well as the formation of new spines and synapses, are key to structural and chemical neuroplasticity that occurs in the microtubules of the subsynaptic zone of dendrites, following learning of experiences [4–7]. The microtubule changes following learning and memory, also form the cornerstone of the neural correlate of perception or consciousness. This implies that the consequence of changes in the microtubules of the subsynaptic zone, following learning and memory of a new experience, will ensure (induce) the emergence of a similar dynamic change in the perceptual state of the brain depending upon the stimulus, through the "Hebbian neuronal assemblies" [38]. In other words, Hebb [16] postulated that functional units corresponding to particular mental states, for instance perception or consciousness might be formed. The functional units are considered as networks or assemblies of many thousands of neurons that can be ignited by particular inputs, and fire in synchrony, and remain active (firing) for hundreds of milliseconds, following which another related assembly ignites, then another, and so on in a phase sequence. Hebb suggested that if the activity of a neural assembly representing a specific set of content exceeds the activity of all other assemblies in some type of competition, it meets the cost of entering into perception or consciousness. In other words, the particular neuron assembly becomes conscious [38].

Several investigators have hypothesized that consciousness is located in a horizontal layer of interconnected cortical neurons sandwiched between ascending (bottom-up) inputs from the thalamus and basal forebrain, and descending (top-down) executive functions from the prefrontal cortex [38]. Bottom-up inputs convey sensory information for general arousal through the reticular formation, as well as emotional context from basal forebrain inputs [106, 107]. Top-down influences include executive functions. Acting together, bottom-up and top-down activations select a neural assembly—a specific subset of cortical–cortical projections—for attention and consciousness [38]. Thus, consciousness is the output of a process that compares available incoming (bottom-up) information against anticipatory executive (top-down) schemata [38, 108].

It is noted that the basal forebrain neurons receive sensory (input) collateral branches from axons en route to the cerebral cortex. In this manner the basal forebrain neurons obtain a general idea of the overall pattern of incoming inputs and then base their outputs on those assessments. The basal forebrain neurons use neurotransmitters acetylcholine and monoamines (norepinephrine, serotonin, and dopamine) [15]. The cholinergic (acetylcholine) neurons contribute to selective attention, whereas the neurons producing norepinephrine, serotonin or dopamine contribute to general arousal of the brain or vigilance. Cholinergic neurons from the basal forebrain are able to contribute to selective attention to a particular sensory stimulus, to a memory, or to a complex amalgam of both stimulus and memory, because they innervate discrete areas of the cortex of a single modality measuring 1–2 mm^2. In contrast, the monoamine neurons innervate the cortex less restrictively, i.e., a single noradrenergic (norepinephrine) fiber may provide inputs to many different types of cortex [15].

Taking into consideration that attention is a deliberate action related to a conscious state of mind, cholinergic and monoaminergic inputs diffusely distributed to cortical modules or to the entire cortex, will sample the overall pattern of input en route to the cortex. These inputs are necessary but not sufficient for attention and arousal. Activation by the neurotransmitters glutamate, acetylcholine and monoamines, as well as the biophysical state of the microtubule network (because microtubules can polymerize and depolymerize, they can search for and activate particular subsynaptic sites) are necessary for full conscious awareness. Consequently, the physical sensation of attention and perception or consciousness are not associated with synaptic activity per se. Instead these sensations are associated with the biophysical state of microtubules that regulates polymerization–depolymerization cycles of their building protein tubulin [15].

Thus, the brain is structured as a consequence of the information stimulus. This takes the form of neural events in space and time that correspond to the spatiotemporal relationships implicit in the stimulus. Areas of the brain are excited and others suppressed [58]. Accordingly, the learning–memory abilities of the brain, as well as, the physical sensation of perception or consciousness are "products" of sensory input–information processing of the brain, depending on the transformations imposed upon the cytoskeleton of neurons, as a consequence of the transformations embodied by the stimulus in the senses [58].

The neurotransmitter glutamate mediates the sensory input to the cerebral cortex, where many of the excitatory presynaptic terminals synapse with the dendrite spines [15]. The number of spines and synapses are more abundant in highly arborized neurons, such as pyramidal neurons, suggesting an augmented excitatory input on the pyramidal neurons, which in turn may mediate motor behavior [3, 15].

Glutamate binds to AMPA, NMDA and kainate, receptors located on the membrane of spines, and opens sodium (Na^+) or calcium (Ca^{2+}) ion channels depending on the receptor, so that Na^+ or Ca^{2+} enters the cell and penetrates the subsynaptic zone. The ionic influx activates proteins which initiate polymerization–depolymerization cycles of tubulin protein of microtubules, resulting in changes in the conformational state of tubulin, which is critical for effective transport of proteins along microtubules to synapses for their strengthening in conveying the information input [15]. Thus, tubulin protein, the building protein of microtubules in the subsynaptic zone of dendrites, has a clear involvement in cognitive processes, which suggests that learning experience alters the basic structure of the cytoskeleton of neurons. Information of the learned experience is then stored in microtubules as memory, which might or might not reach perception or consciousness [15, 38, 109, 110]. In this regard Alzheimer's disease, which is accompanied by deficits in intellect, memory and consciousness, has been linked to microtubule degradation [111].

It has been theorized that the interior regions of the microtubules play the role of electromagnetic cavities. The interaction of these cavities with the tubulin binding protein can lead to the emergence of quantum effects, which play a biologically relevant role in the information processing, storage and retrieval done by the neurons [38].

In this context, the interior milieu of microtubules has been linked to a caged quantum bit (qubit). Conventional computers represent digital information as binary bits of either 1 or 0. Quantum computers can represent quantum information as superpositions of both 1 and 0 known as qubits, and are capable of quantum computation. The conformation states of individual tubulins play the role of bits in computers. The conformation states of tubulin protein then are considered the essential input–output function of the neuron [33].

Accordingly, a large number of authors have provided various arguments for microtubules being pivotal to consciousness. In particular, because they may well form the central nervous system of the cell, since microtubules do propagate signals in cells. It has thus been suggested that if one assumes that the phenomenon of consciousness depends on cellular processes, then nature has provided the necessary structures (microtubules) to operate as the basic substrate of quantum computation, provided there are indeed quantum effects in consciousness [33]. Bohm and Hiley [112] proposed that the wave function in the quantum superpositions (particles/systems existing in multiple states or locations simultaneously, governed by a quantum wave function) contains active information that guides the movement of particles along the microtubules, and that consciousness is associated with this movement and active information. The quantum superimposition state in the microtubules is considered to collapse at the supramolecular level due to interaction with the classical environment or the conscious observer [38].

In this view, the currently most accepted theory of consciousness, the Penrose-Hameroff "Orch-OR" model [38, 113] suggests that our brains are quantum mechanical in their processing of information and that the process of human consciousness cannot be simulated classically. They then proposed that consciousness is a sequence of quantum computations of information in the microtubules within brain neurons, shielded from decoherence to reach the threshold for objective reduction ("OR"). The quantum computations are "orchestrated" (hence, "Orch-OR") by neural/synaptic inputs and extend throughout cortex by tunneling through window-like gap junctions between neurons. Consciousness emerges in dendrites of cortical neurons interconnected by gap junctions (electrical synapses) forming a giant neuron or "hyperneuron," in which all neurons fire in synchrony (simultaneously).

Gap junctions or electrical synapses occur between neural dendrites, between axons and axons, between neurons and glia, between glia, and between axons and dendrites —bypassing chemical synapses [114, 115]. Ions, nutrients and strands of DNA or RNA pass through the open gaps, so gap junction-connected neurons have both continuous membrane surfaces and cytoplasm, and the neurons are electrically coupled, depolarize synchronously and behave like one giant neuron (hyperneuron) [116].

Chemical synapses and axonal action potentials convey inputs to, and outputs from, conscious processes elaborated in hyperneurons. The experiential action potential from the peripheral receptors reaches the cortical dendrites and alters the conformation state of tubulin protein of microtubules. The tubulin subunits of microtubules are the fundamental units of information. Computations are then carried out by the tubulin subunits which spread all over the cortex through the gap junctions generating the conscious state of the brain [38, 113].

It has been suggested that preconscious experience or thought exists in terms of multiple quantum states, and the conscious experience is realized when one of the many possible states prevails. This means that the conscious mental events can be assumed to correspond to quantum collapses of superposition states at the level of macroscopic brain activity [38].

The perspective that could be advanced is that the oral senses are "online" with the brain, so their information input and processing is involved in learning and memory storage abilities, at the same time also giving rise to a conscious experience in the form of perception or consciousness. This may occur when a critical level of complexity and interaction is reached between cortical neurons connected by gap junctions. In other words, if the activity of a functional unit (hyperneuron) corresponding to a particular mental state exceeds all other functional units in some type of competition, it meets the cost of entering perception. Hence, perception chooses reality, while most other mental activities, including motor activity are normally nonconscious activities [38].

Gap junction networks (hyperneurons), which may transmit through the gap junction quantum states from neuron to neuron individual proteins or strands of DNA or RNA molecules, have been shown to be widely prevalent in the brain and to mediate synchronous firing of neurons in various frequencies of the electroencephalogram (EEG). Among these the so-called gamma coherent 40-Hz neuronal activity correlates best with consciousness [117–119]. Clinical signs of pain awareness, pupillary size, heart rate, blood pressure, lacrimation, diaphoresis, mucus secretion, EEG power spectrum/40 Hz, etc, are used to indicate adequate anesthesia and lack of consciousness. While consciousness is erased during general anesthesia, nonconscious brain EEG and evoked potentials continue, although they are reduced [38].

The conclusion seem to be that the information processing in the brain and the resulting structural and chemical changes of the brain can describe the contents of the black box. The brain is "online" with the mouth continually receiving and processing messages from the senses. The information input reaches the microtubules of the subsynaptic zone of the dendrites, that give rise to spines, and induces transformations of the protein tubulin. The resulting changes in the brain can be seen as forming a continuum, i.e., a series of complex correlated mental activities, depending upon the stimulus, which underlies somatic sensation-perception, attention and awareness (arousal of the brain), and learning and memory (cognitive functions), as well as motor output models to muscles through the pyramidal and extrapyramidal tracts [15, 38, 58].

During growth and development gap junctions link cortical pyramidal neurons into functional units (hyperneurons) emitting outputs to the motor nuclei of the trigeminal nerve, resulting in oral sensorimotor functions, such as speech and chewing. A single pyramidal neuron has numerous gap junction connections, of which only some are open at any one time, with rapid openings and closings regulated by the microtubules through their information-processing transformations [120]. This suggests that gap junctions are as dynamic and mutable as those crafted by chemical synapses [121]. Thus within each cerebral hemisphere there is no apparent limit to the extent of gap junction networks, which may form a large continuous neuron syncytium that could even extend to both hemispheres [122, 123].

In this view, thalamocortical neurons generating synchronous alpha and theta cortical activity in the EEG, are linked to gap junctions in the thalamus [123], so that thalamocortical projections (or trans-corpus callosum pathways) could join the two cerebral hemispheres in hyperneurons to account for the bilateral synchronous contraction of the masticatory muscles [38]. Accordingly, abnormal coupling of the cerebral hemispheres in hyperneurons might underlie the dissociation of the left and right masticatory muscle activity in the unilateral posterior cross-bite malocclusion of the teeth [124, 125], suggesting that the dissociation of bilateral activity of the masticatory muscles in the unilateral posterior cross-bite malocclusion of the teeth is due to abnormal brain function.

In this context, the diagnosis of sensorimotor dysfunction of malocclusions of the teeth associated with oral facial imbalances through the information-processing approach is a far different skill from the classical approach. This is because it enables the orthodontist to view the patient as an active processor of information and hence mediate between the patient and the dysfunction by inducing brain neuroplasticity through the orthodontic corrective experience. This implies that neuroplasticity of the brain is responsible for the changes in motor behavior seen in malocclusions of the teeth following orthodontic treatment. Indeed, neuroscientists widely accept that cognition and perception are correlated with the physiological function of the brain [33, 85].

Unlike the traditional orthodontic diagnostic approach, which views only the peripheral behavioral products of the central nervous system, the application of the information-processing model to oral sensorimotor behavior focuses a special interest on the neural events that are necessary for giving an oral conscious experience and how it may regulate oral motor behavior. Thus, the information-processing approach stresses a complete new conceptualization of oral functions through the description of the precise neural events in the brain that underlie these functions, which eventually will provide practitioners with a rationale for correcting dysfunction of the mouth. For example, employment of the information-processing approach allows the practitioner to compare the ways in which different aspects of oral behavior are structured in memory, so that common properties can be discussed. Also, the practitioner who understands the processes that underlie motor behavior is aware of the patient's probable cognitive strategies in situations of discomfort or dysfunction and may contribute to diagnose the source of perceptual-motor deficits. Finally, it is obvious that if orthodontics and dental medicine are to advance, there needs to be a better understanding of the executive centers of the brain, such as the prefrontal cortex, whose output in the form of conscious action focuses on behavior including the oral–facial region.

4.3 Looking into Conscious Brain – the Prefrontal Cortex

The prefrontal cortex seeks continuous sensory stimulation to maintain arousal for attention and awareness. Arousal of the brain guides our thoughts and behavior through strengthening of synapses resulting in enhanced communication and processing of information which in turn induces the conscious state of mind and the control of muscles. Conscious actions and/or consciousness are key to successful life. We need to have clearly defined goals. When we know what we want, we are more likely to change our behavior and the prefrontal cortex function to get it [199].

The prefrontal cortex is the most evolved part of the brain. It occupies the front third of the brain, underneath the forehead. It is often divided into three sections: the dorsal lateral gyrus, the inferior orbital gyrus, and the cingulate gyrus. The dorsal lateral gyrus and the inferior orbital gyrus are often termed the executive control center of the brain dealing with output in the form of action which is involved in the active roles of decision making and "working memory" in a space–time representation of our "sense of future" and of our will or intent. Overall, the prefrontal cortex is the part of the brain that watches, supervises, guides, directs, and focuses the behavior. The prefrontal cortex also helps to solve problems and see ahead of a situation on the basis of experience. This is also the part of the brain that helps in learning from mistakes. People who have trouble learning from experience tend to have poor prefrontal cortex function. They tend to make repetitive mistakes. Their actions then are based not on experience, but rather on the moment, and on their immediate wants and needs.

The prefrontal cortex (especially the dorsolateral prefrontal cortex) is also involved with sustaining attention span. It helps focus on important information, while filtering out less significant thoughts and sensations. Attention span is required for short-term memory and learning. The prefrontal cortex actually sends signals to the limbic system and sensory areas of the brain when you need to focus, and decreases the distracting input from other brain areas [199].

People with attention deficit disorder (ADD) have neurological dysfunction in the prefrontal cortex characterized by a short attention span, which is the hallmark of this disorder. People with ADD have trouble sustaining attention and effort over a prolonged period of time. Distractability, difficulty in learning from past errors, hyperactivity, etc, are some of the problems of ADD. Children with ADD have longstanding problems paying attention in school, and to regular routine everyday matters. When they try to concentrate the activity of their prefrontal cortex decreases, rather than increases, as in normal behavior. When the prefrontal cortex is underactive (neurotransmitter abnormalities may decrease prefrontal cortex activity), as in ADD, too many sensory stimuli bombard the brain and the cerebral cortex is suffering of sensory overload leading to sensory deprivation (see Chap. 3).

Clinical neuroimaging studies of the prefrontal cortex, using for example single photon emission computed tomography (SPECT) which measures cerebral blood flow and metabolic activity patterns, often involve two scans, one in a resting state and a second during a concentration task. In evaluating brain function, it is important to look at a working brain. When the normal brain is challenged by a concentration task, such as mathematical problems or sorting cards, the prefrontal cortex activity increases. In certain brain conditions, such as ADD and schizophrenia, prefrontal cortex activity decreases in response to an intellectual challenge. Integration between the ascending (bottom-up) sensory information from the thalamus and the descending (top-down) executive function of the prefrontal cortex [106, 107] may be affected by neurotransmitter abnormalities [199]. On the other hand, experimental stimulation of the orbital gyrus is known to induce rhythmic opening and closing movements of the jaw resembling masticatory movements [197]. It is then hypothesized that oral neuromuscular dysfunction during chewing includes prefrontal cortex dysfunction through neurotransmitter abnormalities affecting the bottom-up and top-down sensorimotor integration function.

4.4 Programming the Oral Senses for Learning

The effectiveness of an environmental stimulus for processing information in the cerebral cortex is not governed by the volume of the sensory input or the direction of change, but by the amplitude and the rate of change of the stimulus [71]. The sensory receptors respond to the perceptual environment by incorporating ordered and variable environmental stimuli into a personal perceptual scheme. The process consists of extracting useful information from the welter of "noisy" sensations with which the organism or mouth is constantly bombarded [69]. The absence of order (meaningful) or variation in the stimuli tends to degrade the cortical perceptual mechanisms [82]. Perception is always selective and is always testing the sensory modalities that are essential for the maintenance of the scheme of the mouth. With breakdown of this internal frame of reference it becomes increasingly difficult and in fact impossible to structure the environment, and to impose constancy and stability of the perceived natural world. As a result, contours fluctuate, ambiguous figures alternate rapidly, and simple geometric figures change in size and shape [82].

The newborn infant, however, has not yet learned which of the incoming environmental stimuli are meaningful for survival and well-being and which are not [82]. The environment sends information signals in the form of concepts or universals (to use Aristotelian language). If the oral senses are to acquire meaning and become incorporated into the perceptual scheme of the mouth, they must develop models of programming allowing the senses to match to the significance of events that occur in the environment, through learning procedures [69].

In this view, the perceptual organization development of the organism or mouth consists of the capacity to utilize the senses to extract information from the environment in order to build the brain. Without such prior learning of models of the environment, the centrifugal control of afferent sources is without a program, without a basis for predicting that certain events are more likely than others or preclude others (inhibition), and have no basis for selectivity toward stimuli [69].

The prediction of the brain that certain events are more likely to occur than others is based upon several considerations about the nature of perceptual development. The widely accepted view is that continued contact with a rich sensory environment permits the development of differentiation of spheres of activity of sensory modalities, of events within modalities [69]. Early sensory deprivation, prevents such differentiation, and prevents the development of selective gating and construction of models of the environment. It also prevents the development of efficient strategies in motor behavior, since the appropriate responses have not being learned. Behavior development depends on cognitive growth, which consists of learning how to handle the transformations that happen in the brain, following information input [15, 102].

4.5 Oral Learning and Memory of Experiences

The brain changes whenever an animal learns a new behavior, and it is assumed that the change in the brain is responsible for the change in behavior [15]. Once a change has occurred, however, it becomes important to identify the neural substrate that is responsible for the recently acquired adaptive behavior.

The traditional view is that the synapse is the only computationally relevant apparatus in the neuron, i.e., learning occurs in the synapses between the neurons, and the information is stored there as memory [16]. In this view Hebb [16] first proposed that synapses might be reorganized (strengthened) with learning and memory. Hebb suggested that the more a synapse is used, the stronger and more efficient at transferring information it becomes. This enhances neuron communication and information input, which results in learning or enrichment of experiences. Use in turn is driven by experience. Our experiences use specific circuits and synapses, thereby strengthening those very synapses.

Subsequently there have been hundreds of studies examining brain tissue for synaptic reorganization following learning or following comparable physiological stimulation that might resemble learning of neural events. For example, Bliss and Lomo [17] performed experiments in rabbits. They stimulated with a strong repetitive stimulus or tetanus the bundle of nerve fibers entering the hippocampus, a memory center of the brain. The fibers of the bundle form synaptic connections with a special group of target neurons inside the hippocampus. Bliss and Lomo recorded the electrical responses from the target neurons and observed that the shock stimulation magnified significantly the electrical responses from the target neurons, compared to normal responses to a low frequency stimulus. In neurophysiological terms the target responses were potentiated by the test shock stimulus. Bliss and Lomo concluded that increased stimulation for seconds results in dramatic strengthening of synaptic communication, neuron responsiveness, and speed of response that persists for six hours after the increased stimulation has stopped.

Thus, a brief electrical experience lasting some seconds on the presynaptic fibers caused long-lasting responses in the hippocampal memory neurons. The increased stimulation of the neuronal pathway entering the hippocampus resulted in greater synaptic strength, which heightened the function of the pathway in transferring information, as well as accelerated the information processing in the hippocampal neurons, which stored memory of the learned experience. The incorporated memory in the hippocampal neurons allowed them to respond for hours after the stimulus had stopped. In this dynamic view, the memory changes do not consist simply of storage of information. Memory incorporated by experience into the brain alters brain function. Specifically, Bliss and Lomo [17] discovered that in the hippocampus there are special neuronal circuits serving memory, which can show long-term potentiation (LTP). This means that the specialized neurons have the capacity to respond to brief experiences and retain through their memory the firing pattern for long time after the stimulus has stopped [2, 17].

While oral experience increases the strength of synapses, resulting in heightening the flow and processing of information in the brain, on the contrary idleness of the mouth decreases the strength of synapses and impairs learning and memory ability. Accordingly, the LTP has its mirror image, long-term depression (LTD) of oral neural activity [2]. This means that the synapses may undergo graded strengthening or weakening depending on their use (exercise). The brain continually monitors experiences and use of synapses and adjusts accordingly their strength in conveying action potentials. The brain emerges as an ever-growing changing structure that builds its structure and function through learning of experiences [2].

In order to induce LTP, a firing intensity threshold must be reached in the hippocampus memory neurons. This can be done by stimulating a single input neural fiber with a strong stimulus, or many input fibers with a lower intensity stimulus. This implies that weak inputs from many sensory modalities, converging on the same neuron can be bound together in the memory neurons and induce LTP [2]. For example, weak stimulation of the taste fibers and weak stimulation of the visual fibers carrying a representation of an apple could associate the taste and the sight of the apple in memory through LTP of hippocampus neurons.

The molecule that senses that the communicating taste and visual neurons are firing simultaneously is the neurotransmitter glutamate, which is responsible for relaying sensory information throughout the entire central nervous system [15]. Glutamate is an excitatory amino acid released into the synapse in response to an action potential of the transmitting neuron. Glutamate stimulates electrically the signal receiving postsynaptic neuron. Glutamate excites the postsynaptic neuron by binding to special receptors on its surface membrane. The receptor is called NMDA (*N*-methyl-D-aspartate). NMDA receptor is specialized for inducing LTP and storage of memory in the synapses through their strengthening [2]. Experimental data, however, do not consistently indicate that changes in synapses following learning are stable. Thus, a new paradigm may be needed to replace the long-held notion of Hebb's [16] view that information storage occurs in the synapses.

The widely accepted new learning and memory model advanced by Penrose and Hameroff [113] suggests that information storage as memory occurs in microtubules of the subsynaptic zone of dendrites that gives rise to dendrite spines and synapses. In this view, microtubules and MAPs have a clear involvement in learning and memory storage of experiences. Microtubules in the subsynaptic zone are capable of storing important memory information concerning where spines were originally located in the dendrites, and to use that information to determine which spinous synapses will persist or reappear in the event of dynamic retraction following learning a new experience [15, 33, 38].

According to the new learning–memory model the binding of glutamate to NMDA receptors on the surface of the signal-receiving postsynaptic neuron will open calcium channels permitting the electrically charged calcium ion (Ca^{2+}) to enter the neuron. Once inside the cell, Ca^{2+} acts as a signal that stimulates a number of calcium-activated proteins, which would favor breakdown and then rebuilding of glutamate synaptic sites on dendrite spines through polymerization and depolymerization cycles of microtubules in the subsynaptic zone. Hence NMDA receptor is key to neuroplasticity [15].

Specifically, the Penrose and Hameroff learning and memory model suggests that conformational changes in tubulin (the building protein of microtubules) occur when the sensory input reaches the dendrites and is propagated along the length of microtubule cytoskeleton. This means that the sensory information alters the basic cytoskeletal structure of dendrites and as a consequence alters any electromagnetic fields generated inside the microtubules as a result of alteration of transported proteins along them. Microtubules then in the subsynaptic zone of dendrites are good candidates for permanently encoding memory, rather than the synapses [15, 33, 38, 113].

The subsynaptic zone of dendrites gives rise to spines, and many researchers have found experimentally that changes in the size and shape of the spines play a role in encoding memory which is key to neuroplasticity following learning of new experiences [39–41]. Thus, spines and their synapses are not permanent structures. Spines are generally thought to be in a constant state of flux, and this is regulated by various neurotransmitter receptors. Glutamate receptors play a prominent role. High concentration of glutamate produce retraction of dendrite spines. The retracted spines, however, might reappear often in their original location [15].

Microtubules of the neuron cytoskeleton, however, might also be the site of perception and/or consciousness, which form our reality of the world [113]. Our reality is basically our memory. We believe that our existence is so continuous, so seamless, but in reality it is only half a second long. Everything that happened a second before is memory, and the next second—what we believe will happen—is also memory. Thus, the present is a bridge between the memory of the past and the memory of the future [35].

The specialized NMDA receptors on the neuron surface membrane that bind with glutamate transmitting sensory information, as well as the microtubules, are keys in the breakdown and rebuilding of the subsynaptic zone of dendrites, the site of learning, memory, and consciousness [15, 38].

During growth and development the precision of neuron to neuron connection is thought to depend on firing patterns. Neurons that fire together are wired together [2]. NMDA receptors appear to participate in the formation of chemical synapses during development, as well as in the strengthening of synapses in the adult [2]. Chemical synapses using NMDA receptors are critical to perception and consciousness because they are involved in developmental plasticity and learning [126, 127]. There is a family of NMDA receptors in which each member is made up of slightly different molecular building blocks. The different molecular structures, however, change the timing of activation of the receptor by the glutamate, i.e., different family members have different memories. Thus, after glutamate binding, the NMDA receptor remembers and detects coincident presynaptic and postsynaptic neuron firing over a short- or long-term memory [2]. As a result the special neural circuits in the hippocampus serving memory can be potentiated for a short or long time in response to learning an experience, and retain through memory the firing pattern of neurons for short or long time after the experience stimulus has stopped [2].

These temporal memory characteristics of NMDA receptors allow different family members to serve different learning and memory functions during growth and development. Thus, in very general terms the developing embryonic brain is characterized by

NMDA receptors with long memories. With maturity, receptors switch to shorter time sensitivity of memory [2]. The long-lasting sensitivity of neurons of the embryonic brain due to binding of glutamate to NMDA receptors with long-term sensitivity might be the background of the qualitative differences in the oral sensory experiences manifested in childhood between premature and term infants. For example, Bosma [128] suggested that the child who had been premature is excessively sensitive to stimulation of the oral mucosa, and is less able than the term child to make an adequate reflex adjustment to oral stimuli.

This residual aberration of oral sensorimotor control following prematurity may be appropriate to the concepts of neurological maturation as a process of selective diminution of sensory inputs to the brain by acquisition of inhibitory nerve fibers and inhibitory control mechanisms. Accordingly, the child who had been premature may show an excessive and inordinate oral reaction to experience stimuli, due to lack of central inhibitory mechanisms that inhibit oral reflexes and regulate the sensorimotor reaction and hence the learning ability of mouth [128].

Woolf [15] suggested that increased transport of protein NMDA receptor in microtubules of dendrites is critical to learning and plasticity. Black [2] speculated that the switch from long to short NMDA memory receptors might explain why critical periods of learning occur during development. For example, visual deprivation during a restricted period in early infancy results in permanent blindness. Language learning also is optimal during the first six years of life and decreases radically thereafter [102].

Sessle [129] suggested that mastication can be viewed as a learned motor behavior. What is the critical period for learning the chewing behavior? It could be hypothesized that the critical period for learning to coordinate the movements of the mouth during chewing is similar to that of language learning. Bishop [130] suggested that the masticatory muscles during speech and chewing use the same motor units in these two very different types of motor activity. The frequency of discharge, however, of the motor units during speech is far faster than can ever be initiated during the voluntary activity of chewing and swallowing. Accordingly, Goldberg [131] speculated that independent motor programs might exist for the chewing and language functions, associated with independent feedback, such that they both go on at the same time. Thus, the critical periods of learning to speak and to chew might be similar. Actually, during the first year the average child learns and pronounces its first word "Mama" [132]. At about the same time the child switches its feeding pattern, from liquid and semisolid food to solid food, and it has enough teeth to begin mastication.

Learning, however, requires attention. Attention is a deliberate action whereby one directs conscious actions to a particular sensory stimulus, to a memory, or to a complex amalgam of stimulus and memory. Taking into consideration that the child's mind is in a constant state of flux (possibly because dendrite spines are in a constant state of flux regulated by various neurotransmitter receptors), the sensory inputs often do not reach perception or consciousness [15]. Consciousness and memory, however, are interrelated psychological phenomena [15]. This means that a conscious observation is by definition an observation that is stored in memory [133]. Thus, in order to make pragmatic use of information stored in memory, that information must be brought to consciousness. Psychologists refer to this process as memory retrieval [15]. Memory retrieval might be difficult in children due to the constant state of flux of their mind. Accordingly learning procedures in children might require increased attention.

In sum, the cognitive approach to motor behavior stresses a complete conceptualization of behavior through the role of information processing and attention in learning of sensorimotor activities. This awareness increases the capacity of the practitioner to diagnose and treat oral motor deficits through conscious reinforcement of motor function learning procedures associated with memory storage.

4.6 Looking into Memory – the Temporal Lobes

For too many years the temporal lobes have largely gone unnoticed in human psychology. They are rarely discussed in psychiatric circles, and few neurologists have been concerned with the rich contribution they make to who we are and how we experience life. Until clinical neuroscientists were able to map activity in the temporal lobes, their function remained mysterious. Many professionals basically thought of them as arm-rests for the brain. The brain imaging work clearly shows that the temporal lobes play an integral part in memory, emotional stability, learning, and socialization. The most precious treasures we have in life are the images we have in the memory banks of our brains. The sum of these stored experiences is responsible for our sense of personal identity and our sense of connectedness to those around us [199].

On the dominant side of the brain (the left hemisphere for most people), the temporal lobe is intimately involved with understanding and processing language, intermediate- and long-term memory, complex memories, the retrieval of language or words, emotional stability, and visual and auditory processing.

The nondominant temporal lobe (usually the right) is involved with recognizing familiar faces and facial expressions, being able to accurately perceive voice tones and intonations and give them appropriate meaning, hearing rhythms, appreciating music, and visual learning. The temporal lobes have been called the "interpretive cortex" as they interpret what we hear and integrate it with stored memories to give meaning to the incoming information [199].

4.7 Conscious Cognitive Aspects of Oral Motor Behavior

In early postnatal growth and development period, during which the infant learns the functions of biting and chewing, which substitute the inborn suckling behavior, it is interesting to examine the mental events that precede oral motor behavior and need to be learned first.

One attempts to understand at first the salient features of somatosensory input that the mouth provides, as well as the visual, auditory and touch input that the mother provides during feeding of the infant. Thus, complex neuronal assemblies in the somatosensory, visual, and auditory cortices respond to these stimuli, through transformations of microtubules and storage of memory in the subsynaptic zone of dendrites [15, 38].

The perspective advanced is that a specific fingerprint defined by a particular electromagnetic state of a microtubule array potentially corresponds to a unique unit of cognition, for instance a basic visual parameter, a sound, the frequency of a tactile stimulus, etc [38]. These learning effects, however, do not explain how the simple perception of a sensory modality of the mouth contributes to the realization and learning of complex motor tasks, such as chewing, through multisensory inputs.

Accordingly, it has been suggested that cross modality influences might occur through feedback connections from multisensory cortices [15]. For example, a recent functional magnetic resonance imaging (fMRI) study has identified a region of the brain which is activated specifically by a combined visual-auditory task [134]. In another study neurons in the secondary auditory cortex tuned to precise frequencies of sound also responded to somatosensory input [135]. Similarly, it has been found that visual stimuli enhance tactile acuity through modulation of responses in the primary and secondary somatosensory cortices [136]. Thus, crossmodal influences, which appear instantly and disappear rapidly, may occur during early development in the primary and secondary sensory cortices, and serve multisensory tasks [15]. It has been suggested that most cognitive tasks we perform have multisensory components [15]. Accordingly, the crossmodal regions of the brain with multisensory connections might be involved in learning. For example, if memory as suggested is stored in the microtubules, and if the microtubules are also the site of perception or consciousness, then it is a simple case to relate memory and consciousness [137].

On the other hand, Libet [138] suggested that a prolonged cortical activity of 500 ms is necessary to produce a conscious experience. This may mean that during the half-second time of cortical activity, a number of unconscious correlated neural events have to occur in microtubules before conscious processes are established. In other words, conscious experience is an experience which is stored in memory [133]. Commenting on this, Gray [139] pointed out that, for instance, in a visual field we are conscious only of visual entity (gestalt), not of visual components (e.g., shape, color, motion, etc.). Hameroff [38] proposed that each visual component is processed in separate brain areas and at different times. The cumulative effect, however, is integrated into a unified visual gestalt with the growth of a hyperneuron, implying that the neurons of all visual components are connected with gap junctions forming a giant neuron or hyperneuron.

In addition, Penrose and Hameroff [113] suggested that unconscious processes capable of becoming conscious utilize quantum information. This means superpositions of both firing and not-firing neurons as quantum bits (qubits) in the microtubules. Qubits are manifested as switches that utilize superpositions at various quantum states. In microtubules, which behave like quantum computers, occurs the transition from nonconscious (superposition quantum information) to conscious (classical) information through decoherence or objective reduction. Decoherence is the disruption of quantum superposition states due to energy or information interaction with the classical environment, or observation by the conscious observer in the space–time geometry.

Accordingly, one might assume that conscious (voluntary) chewing is a multisensory task that might also include crossmodal input information, leading to the emergence of a number of unconscious correlated mental activities, which underlie transformations of separate brain areas. When the activity of neurons corresponding to transformations of separate brain (correlated) areas reaches the "threshold" for synchronous firing, gap junctions appear between the neurons and unite all the correlated components into a giant neuron or hyperneuron for synchronous firing activity, corresponding to conscious identification of the cumulative effect (gestalt) of the correlated mental activities, which in turn are experienced by the individual as sensorimotor abilities, such as biting, chewing, occlusal relationships of the teeth, or tongue movement within the scheme of the oral cavity, etc.

However, since the brain of young children is in a constant state of flux, having thus difficulty in concentration and attention (see Sect. 4.5), a number of children may have difficulty in learning the mature oral sensorimotor activities which require increased attention and awareness, because the sensory input does not result in correlated mental activities and/or the correlated mental activities do not reach conscious states through synchronous firing in hyperneurons.

The biochemical correlates of neurons in changing behavior are sometimes ascribed to "stress." In so far as a natural or experimental situation requires a new response to be learned, a stress is in one sense experienced. A new state must replace the old state of equilibrium in the brain. This is a stress in the same sense that distortion of bone constitutes a stress that leads to structural remodeling—as a result of the remodeling the distortion of the bone is minimized. Thus, both in brain and bone the stress is a localized distortion. Biochemical correlates of such a stress are a proper part of the learning process [140]. In this view, information input to the brain that reaches perception or consciousness might be considered as a "stress" because it calls for learning of new associations of behavior through structural and functional transformations of the brain, i.e., the conscious brain is the learning brain. This may imply that the conscious child might be reluctant to learn processes that are experienced as stress.

Watson [140] also suggested that experience gained at certain critical stages of development of the brain has a persisting influence upon motor behavior. This is because early learning of experience is register in the brain. If certain experiences are not gained within the early growth and development period, but later, the facilitation of normal behavior fails, or else a far greater intensity and duration of stimulation is needed to produce a much smaller effect. This implies that options for determining certain patterns of behavior do not remain open indefinitely, even though these patterns may not emerge until a later stage in the growth and development period. There is evident similarity between this and the dependence of proper development of the visual cortex upon early experience. For example, neurons of the visual cortex produce massive amounts of tubulin (the building protein of microtubules) during critical periods (from the day the eyes open to postnatal day 35). The critical period is the time during which synaptogenesis and visual learning occurs at the highest rates [105].

By the same token, we may assume that there is a critical period during which the infantile innate suckling behavior has to be replaced by the normal biting and chewing functions through conscious learning. This substitution may not be facilitated by the sensory (experiential) input at later stages of development and thus the oral sensorimotor function may not reach the mature pattern possible through hyperneuron formation. Instead the oral behavior remains infantile and stereotyped and is the pattern we encounter in certain malocclusions of the teeth.

Another category of oral sensorimotor immaturity is the influence of complete deprivation of sensory experience of one modality upon the ability of the child to discriminate. If the sensory deprivation occurs throughout a critical period of development, the impairment of discrimination is irreversible or partially reversible only if a careful training program is used [140].

As we come to understand how the brain does or does not incorporate information that arises in the oral senses resulting in conscious learning, we may be able to give some coherent account of the localized cognitive cortical dysfunction in relation to oral sensorimotor dysfunctions we encounter in the malocclusions of the teeth. The changes in the brain following normal information input or through sensory deprivation of the brain are responsible for the development of normal or abnormal motor behavior, respectively, during growth and development and thereafter [15]. Similar brain changes could be induced during orthodontic therapy (see Chap. 10).

Children do not react passively to orthodontic procedures. Instead, they are or should be conscious processors of information input, actively operating to transform the input into cortical neural activity underlying learning of new patterns of motor control of muscles [141]. Gibson [141] and other perception psychologists have emphasized that we should think of the act of perception (and of the learning process) as dynamic, not static. There is a constant stream of stimuli, much like the constant stream of water moving as a soliton in a canal, in which the essential geometric and temporal relations are preserved as an aspect of neural activity. For example, if we imagine the situation of a child who is too young to recognize the "data stream" of a triangle, the same set of transformations is imposed upon the visual cortex as in the case of an older child who is able to recognize the triangle. The brain changes its cognitive state depending upon the stimulus. In the young child who does not recognize the triangle, there is no ordered relationship among the "data stream" visual stimuli, and thus they may be moving through the soliton randomly with respect to one another, so any relationships that may be formed quickly become unstable. However, solitons or traveling wave solutions depend upon symmetries or invariance within the medium and its boundary conditions. It is the order that is implicit in the stimulus, the symmetries or invariance that facilitates the emergence of traveling wave solutions. In these the only stable dynamics that are possible are those that unite the stimulus in terms of its own implicit spatiotemporal invariance. In this aspect, the soliton is an ideal candidate to form the basis of cognition [58].

Accordingly, paraphrasing Mountcastle [62], it could be said that the more specific (patterned) is the sensory input, the more conscious learning states can be reached by the cortical neurons, and then the more specific will be the motor output patterns. This specificity between the information-conscious mind and motor behavior is discussed next.

4.8 Perception Regulates Oral Motor Behavior

Brains of course are involved in many activities but they share a common theme: information processing. The conscious brain processes a vast amount of information concerning the interaction between the body and its environment through a complex network of chemical signaling neural pathways [85].

But brains are not always conscious. The brain remains very active in the state we call unconsciousness, and even when we are conscious, we are only aware of a very small trickle out of the vast quantity of information flowing through the brain. There is clearly a tendency for the more primitive activities—those we share with lower animals—to be performed unconsciously and the more specialist higher level motor activities, such as reading and writing (language use generally), to be accompanied by consciousness because they require the unique computational abilities of the conscious mind to be learned. It is also intriguing that the obligate conscious activities include memorizing information and learning, but recall of memorized information (such as the motor actions required to perform a learned task) does not require consciousness. Thus, any theory of perception or consciousness must clearly delineate the operational difference between conscious and unconscious mind [85].

There is a considerable body of opinion that asserts that consciousness has no impact on the way the brain works. In other words consciousness is an epiphenomenon of the brain's activity. If this is the case then consciousness makes no difference to the world; it generates no facts. In other words there is no scientific basis to deal with. It is not a subject for scientific theories [85]. But consciousness does of course generate phenomena. It impacts upon the world, it is a property of living brains. Brain activity may be conscious or unconscious. The conscious mind appears to be serial. Learning requires consciousness, but recall does not. Consciousness correlates with synchronous firing of neurons and with the gamma frequency of the encephalogram [15, 113].

Specifically, McFadden [85] introduced the conscious electromagnetic field (CEMI) theory which suggests that consciousness is a product of the brain's electromagnetic field. McFadden proposes that the source of the quantum effects in the brain would not be microtubules, but interactions between the brain's electromagnetic energy field and the voltage-gated ion channels in the membrane of the neuron. The ion channels in the membrane are involved in information processing. The opening of membrane ion channels depends on quanta of electromagnetic energy that must be absorbed by the surrounding neuron field to open a single voltage-gated ion channel, which will initiate firing (action potential) of the neuron.

As already discussed, in the Penrose and Hameroff "Orch-OR" [113] model of information processing, the synchronous neuron firing in consciousness is a correlate of attention and awareness which are considered as preconsciousness states of the brain elicited by fast transmission of neural signals in the hyperneuron through gap junctions between the neurons. The synchronous firing of neurons in the hyperneuron creates consciousness. On the contrary, McFadden [85] suggests that electromagnetic energy fields may be involved in mediating and maintaining firing synchrony of neurons and thereby potentially provide a substrate for binding the neurons together forming a conscious state.

In fact, the CEMI theory [85] proposes that when a neuron of the brain receives a signal from a peripheral sensory neuron, synaptic neurotransmitters stimulate ion pumps that cause postsynaptic neuron membranes to become negatively polarized, depending of the type of signal received. If the membrane potential falls below −40 mV, the neuron fires. The massive membrane depolarization of neuron will also generate a conscious electromagnetic energy field perturbation that traveling at the speed of light can integrate the diverse information (computational activity) held in distributed neurons. The layer (laminar) organization of the neurons of the neocortex and hippocampus with parallel arrays of neurons will tend to amplify local electromagnetic energy fields. Thus, the electromagnetic field across a neural membrane will be the product of a field generated by the membrane dynamics (ion pumps) of firing neuron, but also by the fields generated by the resting states and firing of all the other neurons in the vicinity.

The voltage of the electromagnetic energy field across the membrane of the firing neuron will be quite weak. However, in a busy brain, through attention and awareness, it is likely that many neurons will be in that state, so the electromagnetic energy field that the brain's activity generates will inevitably influence neural dynamics, resulting in a conscious electromagnetic energy field, which is the physical substrate of perception and/or consciousness [85].

In this view, information processing (decoding) within each neuron is pooled and integrated to form a unified electromagnetic information field of conscious percepts. In other words, if individual neurons represent features of a scene, or of an object (color, shape, movement, etc), the conscious identification binds together the different features and creates an image of the ensemble, like the pieces of a puzzle, through the synchronous firing of the neurons involved in the different features of the scene [85].

A distinctive characteristic of the brain's conscious electromagnetic field integrated in neurons, is that perception corresponds to only that component of the electromagnetic information field that initiates motor actions. In other words, perception moves the muscles and is thereby capable of communicating its state to the outside world [85]. The site of action of the conscious electromagnetic field is most likely the pyramidal neurons of the cerebral cortex, which in order to induce perception should be busy, i.e., to be involved in a number of correlated perception mental activities, such as, attention, and awareness, characterized by synchronous firing of neurons. Thus, in sending conscious electromagnetic information to muscles through the pyramidal neurons, the brain exerts conscious control with profound functional consequences on motor behavior [85].

An example of synchronous firing of neurons associated with attention and awareness is the following. Kreiter and Singer [143] have demonstrated that neurons of the visual cortex in monkey brain that responded to two independent images of a bar on a screen fired asynchronously when the bars were moving in different directions. The same neurons, however, fired in synchrony when the bars moved together. The awareness of the monkey that the bars represented two aspects of the same object, was encoded in the same visual cortex by the synchronous firing of neurons. Thus, awareness and attention correlates with perception not with a pattern of neuron firing, but with synchrony of firing neurons. It has been suggested that the synchronous firing links distant neurons involved in registering different aspects (color, shape, movement etc) of the same visual perceptions and thereby binds together features of a sensory stimulus [144].

It has been known for long time that concentration upon a mental task can lead to activity in muscles not involved in its performance. The term "stress" or mental stress has been applied and stressful physical tasks in various forms have been shown to result in muscle activity (measured by means of EMG techniques) in many muscles of the body [95]. For example, Perry et al. [145] demonstrated stress-produced hyperactivity in the masseter and temporalis muscles in dental students during a stressful interview. Also the same investigators reported that tooth clenching occurred during a stressful interview. Similarly, Yemm [146] using a task requiring mental and physical coordination, demonstrated several characteristics of the response of the masseter and temporalis muscles placed under stress. It has been suggested that the muscle response is a result of a generalized activation of the gamma motor system from higher centers involving primarily the jaw-closing muscles.

According to the electromagnetic field [85] theory, however, the mental stress might be regarded as a particular brain activity associated with conscious electromagnetic information, which instructs the jaw-closing muscles to contract. For the same reason all verbal thought is accompanied by subvocalizations (i.e., motor cortex activity accompanied by appropriate but normally undetectable vocal tract activity) [147].

In this context, perception may be a neuropsychic state of readiness for mental (attention, awareness, learning, and memory), as well as of physical (motor) activity. This in turn may be the reason that conscious motor activities are always accompanied with attention, awareness, and cognition [15]. The view that perception is a neuropsychic state of readiness for mental and physical activities is supported by Hameroff [38] who suggested that perception is involved in the control of actions, self awareness, feelings, choices, phenomenal experience, a model of the world. Similarly, McFadden's [85] CEMI theory views conscious will as having a deterministic influence on motor actions. This means that perception or consciousness plays a causal role in determining our conscious motor functions—something like mind affecting matter. McFadden suggests that we are not simply automatons that happen to be aware of our actions. In contrast, perception provides a physically active role for "will" in driving our conscious actions.

Accordingly, the CEMI field theory proposes that any voluntary motor activity of the body will require the influence of the conscious electromagnetic information field input in order to be learned. The conscious mind will then push the motor neurons to shift towards some desired motor activity [85]. For example, in order to learn the chewing act, the electromagnetic field will initially be required to provide that fine control of coordinated motor activity necessary to perform the skilled action of mastication. The strengthening of synapses by frequent stimulation of masticatory motor neurons through frequent use of the chewing action will contribute to augmentation of the sensory input and processing followed by augmentation of the conscious electromagnetic field of the motor cortex. After repeated augmentation of sensory input processing function, the movements of the masticatory muscles during eating will become increasingly independent of the conscious electromagnetic field influences. The motor activity of mastication has been "learned" and thereafter may be performed unconsciously without conscious CEMI field influence on the neural networks involved. Hence, a child will be able to chew without conscious input [85].

After the chewing function has been learned the CEMI field, in the absence of any conscious motor output to the masticatory neurons, is involved in strengthening the synapses in the "masticatory pathways" in order to "hard-wire" neurons and thereby lay down long-term memories of the learned masticatory function. This is because the conscious CEMI field is a continuum of values (information) extended in space and time and its activity is required to store memory of the learned motor tasks. Accordingly recall of the memorized chewing action may be unconscious [85].

The voluntary contraction of individual masticatory muscles is largely a function of the corticobulbar tract, while the extrapyramidal tract composed of all pathways descending from the cortex (except the pyramidal system), including the basal ganglia, subthalamus, cerebellum and brainstem, is an activator and coordinator of postural adjustments and of automatic (unconscious) movements of skeletal muscles, including the masticatory muscles [36]. Thus, in terms of the CEMI theory, the learning in childhood of the precise and dexterous movements involved in chewing require the function of the direct corticobulbar motor system, which is substituted after learning has been completed by the indirect automatic (unconscious) extrapyramidal tract. This view, is supported by studies in humans suggesting that in the adult cerebral cortex the masticatory movements are not located in the precentral and postcentral gyri, as are most other movements of the mouth [66]. There are very few if any similar studies in children during the learning masticatory period. It is possible, however, that the masticatory function is represented in the cortex during the voluntary learning of the act, and afterwards when chewing becomes an automatic unconscious function the representation sinks to the subcortical structures.

Perception psychologists have emphasized that conscious learning of motor behavior is always a selective process, which is always testing the sensory modalities. One has to think in terms of a dynamic mental activity requiring much attention and awareness of the individual. During the attentive learning state of the cerebral cortex, the individual strives or seeks continually to find variable stimuli (different stimuli of different intensity), as well as ordered environmental stimuli (symmetries or invariance implicit in the stimulus) that are vital to program its senses and maintain arousal of the cortex for attention and awareness, through the appropriate input from the senses. The cortex in turn will accomplish the learning contributing thus to the physiological efficiency and stability of motor behavior [82, 141]. The absence of variation and order in the environmental stimuli tends to degrade the cortical perceptual organization and learning ability, resulting in impairment of motor behavior [82].

The process of incorporating variable and ordered environmental stimuli into the senses (see Sect. 4.4), representing a personal perceptual scheme of the world, is also important for growth and development of the body. For example, Schanberg [73] recently suggested that rat pups deprived of touch stimulation by their mother experience a significant decline in growth hormone and ornithine decarboxylase (ODC), which is part of protein synthesis. Schanberg located a growth gene that is responsible for this relationship between touch stimulation and growth. Specifically, he identified a messenger gene (CFOS), and he found that maternal touch deprivation significantly reduces this messenger. Schanberg concluded that "the brain (reacting to environment) can stick its long arm down right into the middle of a cell and regulate a gene."

Similarly, Moss [91] considering the implications of mechanosensation on bone growth and remodeling, suggested that mechanosensory systems operate in bone organization and growth. Specifically, Moss suggested that bone cells (except osteoclasts) are extensively interconnected by gap junctions (electrical synapses) forming a large connected cellular network (syncytium). Bone cells are electrically active. Bone cells are stimulated by the contraction of muscles which tends to deform bone (strain) and can initiate membrane action potentials capable of transmission in the entire bone cellular network through the gap junctions. In a very real sense bone tissue is "hard-wired." Skeletal muscle contraction is a typical mechanotransduction process, having morphogenic effects on bone through intraosseous production of electric fields tunneling in the entire bone cellular network through interconnected gap junctions (electrical synapses). Subsequently the ionic signal information is computed in the interconnected cellular network of the bone (probably in the microtubules of osteoblasts) so that organizational adaptive responses of the bone are correlated (tuned) to precise frequencies of muscle contraction. Moss concluded that mechanotransduction and computational bone biology offers an explanatory chain extending from the epigenetic event of muscle contraction, downward to the bone cell genes. Thus, it appears that the information processing of the oral experience in the brain is linked with a series of complex researchable mental processes, such as conscious learning and memory of motor responses to muscles [85] which in turn are involved in bone organization [91] and growth and development processes [73].

In sum, our special interest in the development of masticatory function is focused on the neural events of attention and awareness, which are necessary for giving conscious perception of oral experiences. Conscious perception emerges in the pyramidal motor neurons connected by gap junctions forming hyperneurons, in which the component neurons fire in synchrony [38, 113]. Hyperneurons of the left and right cerebral hemispheres connected by the corpus callosum pathways account for the bilateral synchronous contraction of masticatory muscles [38].

Alternatively conscious perception emerges through the conscious electromagnetic field generated in the pyramidal motor neurons, which provides specific instructions to individual masticatory muscles how to contract [85]. It is hypothesized that the conscious electromagnetic field in turn programs the cyclic aspects of chewing movements controlled by the "central pattern generator" or "chewing center." The chewing center is located in the reticular formation of the midbrain at the cross-roads of the sensory input and motor output systems of the cranial nerves [76]. The ascending reticular activating system (ARAS) regulates the ebb and flow of activity in the afferent and efferent systems of the reticular formation and exploits aspects of the information processing capabilities of the cerebral cortex, such as for instance perception, learning, memory, and motor control of muscles [64]. The ARAS may be involved in the programming of the chewing center by associating the sensory and motor features encoded in the memory of the component neurons, through the influence of the conscious electromagnetic field.

This model of the development of the chewing function stresses the importance of patterns of conscious experience in the development of coordinated masticatory movements by individual central design, through learning and memory. This view is best exemplified by the suggestion of Bosma [43] that the oral innate stereotype functions of a newborn child, such as the suckle function, change postnatally only by the addition of awareness and volition. The conscious electromagnetic field may actively be involved in this change.

4.9 Excitable Media: An Alternative Approach to Brain Structure and Cognition

Davia [58] advanced the hypothesis that the brain is not an ordinary excitable medium in which each part of the medium is isolated and then connected to other parts of the medium with the development of excitable fibers (nerves). The structure of neurons and of the entire brain can be understood as a self-organizing excitable medium, structured by the senses. In this view traveling waves or solitons are associated with self-perpetuating cycles of catalyzed reactions, resulting in conformational change of a protein (change of shape), for instance of the protein tubulin of microtubules, through a "quantum tunneling" mechanism. This is facilitated by a vibrational mode of the protein in the form of a soliton or traveling wave through the excitable medium. Localized nonlinear traveling waves, of which solitons are an example, are known to represent an efficient way of transferring energy in biological processes at different scales (microscopic or macroscopic). The phenomenon of traveling waves is expressed as movement of molecules of a protein, which acting as an enzyme (catalyst) accelerates the reaction between chemical reagents by transferring energy from one place of the protein to another, hence achieving transition to a more favorable thermodynamic state. The individual molecules of protein become united in a continuous coherent traveling wave, which is a quantum unified state of structure/energy.

The excitable medium is a substrate with energy associated with it. Once the energy of the substrate has been depleted as a consequence of a metabolic process (or autocatalytic reaction), there is a system to replenish the energy required. For soliton wave formation there must be symmetry or invariance in the structure of the medium and boundary conditions. For example, microtubules have a regular depth and width forming a tube, like a canal, which is structured by the senses. That symmetry in structure then translates to excitable areas in the brain with the emergence of traveling waves. Thus, the spatiotemporal evolution of solitonic waves or traveling waves of the excitable medium overcome the structural constraint of the medium by providing paths for energy dissipation via the invariance or symmetries that are implicit in the structure of medium [58].

The action potential of nerves is another example of a self-perpetuating autocatalytic cycle that exhibits soliton-like behavior [207]. Similarly, the relationship between solitons or traveling waves and structural regularity is also evident in the dynamic interaction between microfilaments, myosin and actin in muscle contraction. Energy is released from the muscle fiber as a result of the action of traveling waves. The myosin molecule acts as a catalyst protein employing the same mechanism as the enzyme, the soliton, to achieve the transition of energy such that the myosin molecules slide and rebind with the actin molecule and thus decrease the overall length of the sarcomere during contraction [208]. Any small defect in the muscle fiber represents a discontinuity that cannot be integrated into the overall traveling wave function. Such a defect gives rise to irregular muscle contraction rhythms. Similarly, the beating of the heart is associated with a traveling wave of excitation that moves through these tissues. The traveling wave is a continuous phenomenon. The defect in the heart muscle represents a discontinuity that cannot be "integrated" into the overall traveling wave solution, resulting in cardiac arrhythmia. In cardiac arrest an electric shock to the heart may be required to initiate again the beating of the heart according to its own internal dynamics, i.e., by the traveling waves through the heart tissues [58].

The theme of excitable media and traveling waves also plays an important role in the auditory system. The basilar membrane is an excitable medium. Its motion consequent on a sound stimulus is a traveling wave [209]. The basilar membrane is not a passive "receiver" of sound stimuli. Its frequency responses are highly nonlinear and its sensitivity is increased by active feedback processes [210]. Specialized hair bundles have the effect of altering the dynamic response of the basilar membrane thus altering its sensitivity to particular frequencies of stimulus. The dynamic feedback mechanisms associated with the basilar membrane not only respond to a stimulus but also "anticipate" it [211]. This interaction of the basilar membrane with the sound stimulus forms the basis of the development of language and speech which is discussed in the next chapter.

In sum, in the previous sections of this chapter it has been suggested that successful oral behavior is necessarily correlated with robust neural states of the brain that underlie sensation, cognition, and motor control of muscles. Perceiving the brain as an excitable medium may offer further insight into brain function especially how different muscle functions are coordinated. Davia [58] suggested that when we speak we coordinate the movements of lips, tongue, and the shape of the mouth with the sounds that we hear. This synchrony is achieved as a consequence of traveling waves uniting the neural activity in different parts of the medium (brain). This hypothesis may also be applied to the coordination of the masticatory muscles during chewing as follows.

Bishop [130] suggested that the masticatory muscles during speech and chewing use the same motor units in these two very different types of motor activity. However, the frequency of discharge of motor units during speech is far faster than in chewing. Commenting on this, Goldberg [131] suggested that there is no question that there are independent motor programs for speech and chewing activities associated with independent feedback. Recent neurophysiological studies have suggested that skeletal muscles are excitable media. The myosin molecule acts as a catalyst protein and employs the same mechanism as the enzyme, the soliton or traveling wave, to achieve the transition of energy enabling the myosin molecules to slide and rebind with the actin molecules and thus decrease the overall length of the sarcomere during muscle contraction [208].

Accordingly, it could be speculated that the paired masticatory muscles are excitable media, each pair forming an independent sensory–motor program associated (correlated) with independent neural states (transformations) of the brain. The neural states are in turn interconnected with traveling waves uniting the neural activities in the different parts of the excitable medium (brain), thus providing temporal synchrony in their discharge and coordination of the frequency output to the masticatory muscles. Thalamocortical projections (or trans-corpus callosum pathways) could couple the two cerebral hemispheres in hyperneurons to account for the bilateral synchronous contraction of the masticatory muscles [38] (see Sect. 4.2).

The contraction of muscle is associated with a traveling wave of excitation that moves through the tissues. The traveling wave is a continuous phenomenon. A small defect in the excitable medium (muscle) represents a discontinuity that cannot be "integrated" into the overall traveling wave function. The consequence of such defects is to give rise to irregular muscle contraction rhythms [58] that we often encounter in malocclusions of the teeth.

The coordinated brain output to the masticatory muscles may further be connected through traveling waves with the neural activity of the basilar membrane, which is also an excitable medium. Its motion responding at the frequency of a sound stimulus is a traveling wave [209]. Thus, we could coordinate the masticatory movements with the sounds we hear when we chew. Similarly, when we speak, we coordinate the movement of our lips and tongue and the shape of our mouths with the sounds we hear. Davia [58] suggested that this synchrony is achieved as a consequence of traveling waves uniting the neural activity in different parts of the brain.

5 Language and Speech

5.1 Introduction

Lines of evidence indicate that there are no innate language representations in the cortex, and that regions of the cerebral cortex that normally support language can support other functions. On the other hand high-density event-related potential (HD-ERP) studies in infants have suggested that some regions of the cortex, such as the left temporal lobe, may be particularly efficient at processing speech input from the first few months of life. These findings have sometimes been characterized as evidence of "innate language neural module," but they are also consistent with a more probabilistic epigenetic view of language function [102]. Accordingly, the view that holds is that speech and language are learned sensorimotor functions. This means that the child does not progress from crying, to babbling, to talking without help. The child must be taught to talk, and the parents are the child's first teachers [132, 214].

Speech is amplified and filtered in the mouth that provides a resonating cavity that helps shape the sound for speech or song. Almost any intervention that a dentist, orthodontist or prosthodontist undertakes in the oral cavity of the patient can have varying degrees of temporary or permanent effects on oral proprioceptive perception and oral motor control, and therefore speech production. Most people, however, adapt surprisingly well to structural modifications of their oral cavity.

On the other hand, patients may seek help from their orthodontist in order to improve their speech, if the speech problem is related to dentition or occlusion of the teeth. A basic understanding of how language and speech development is accomplished might be useful to practitioners working in the mouth.

5.2 Language and Speech Development

A few months after birth the infant is engaged in considerable amount of oral activity. For example, from the age of 5 months everything the infant picks up is taken into the mouth. This mouth and hand experience can provide stimulation to all senses with which the baby

M. Z. Pimenidis, *The Neurobiology of Orthodontics*,
DOI: 10.1007/978-3-642-00396-7_5,

introduces himself to the world. The mouth and hand sensory experience (mouthing) persists until the baby is really adept with his hands [70]. At 6 months the oral cavity has grown considerably and the tongue no longer fills the whole mouth. The larynx starts to sink down in the neck and the velum or soft palate now can be elevated to close the velopharyngeal sphincter, which regulates the nasality of the speech sound [148].

Speech is generated in the oral-pharyngeal cavity, which filters the rough sound signal produced in the vocal cords. In most people the air is amplified and filtered in the mouth that provides a resonating cavity that helps shape the sound for speech or song. For these sounds the velum must be elevated to close the velopharyngeal sphincter so that no air goes through the nose [148]. Thus, the low position of the larynx in the human neck, the control of the velopharyngeal sphincter mechanism, and most important the preprogrammed brain to develop language if there is linguistic input in the environment, are the main special features that predispose for speech production [102, 148].

The central functions of language comprise the cognitive functions and the motor aspects of vocalization. The cognitive functions involve the processing of linguistic input in Wernicke's cortical area of the temporal lobe of the left cerebral hemisphere, which comprises the center of semantic meaning of words or the unspoken word function center. The ability to produce spoken words (linguistic articulation center) is located in the premotor area of the same hemisphere, specifically in the areas 44 and 45 of the inferior frontal gyrus known as Broca's area. Isolated lesions of Broca's area result in motor aphasia, demonstrated as motor apraxia, in other words, loss of articulate speech but without demonstrable paralysis of muscles associated with speech [36].

At birth the cerebral cortex is an "unwritten" page. The eyes and occipital lobes are not yet developed. The eyes are open but the infant cannot see, hear, smell, or touch. Thus, the child's sphere of sensations is limited at this early age. These acts require the activity of definite groups of muscles, which cannot be controlled by the newborn child. For example, to see an object it is essential that the optical axis of both eyes converge on the object. This can only be effected with the help of the muscles, which turn the eyes in all directions. The newborn child cannot do this. Its look is always uncertain, i.e., it is not fixed on definite objects. After one or two months the child learns to perform these movements through the sensations that impart to the cerebral cortex the musculovisual reflex association. At the same time the neurons of the optic tracts are ready to transmit the impulses from the retina to the occipital lobes so that the infant can see the mother [68].

Similarly, a long time passes before the infant learns to hear sounds and words (listen). This prelistening stage of development is expressed as beginning to babble. In other words, the infant is babbling, with coordinated movements of the lips and mandible, producing simple repetitive consonant-vowel sequence sounds (mamama … bababa) without meaning. Up to this stage a baby's vocalizations are largely unrelated to its race or hearing [148]. But soon the factor of imitation of learned sounds appears, through memory development, usually between six and ten months of age [68].

In the listening stage of speech development the brain is in an attentive state during which learning occurs [15]. A decisive role is played by the child's desire to imitate the articulation of sounds and their combination into words, acting on its eardrum. The process of articulation of sounds consists of association of sensations caused by the contraction of muscles participating in speech with auditory sensations induced by the sounds of the individual's own speech (babbling). These acts are reflexes arising in the brain, through memory stimulation, and end in muscular movements. These reflex sensations are acquired by learning through frequent repetition, and their reproduction requires memory [15, 68].

The act of auditory attention or listening is similar to the described convergence of the optical axes on the object for clear vision. Listening is manifestly confined to this external act when separate simple sounds are perceived. In the vocabulary of a child there is not a single word that it has not been acquired by learning.

The process of articulating sounds is actually the same in the child and in the parrot. But what is the difference in their faculty of speech? While the parrot learns to pronounce only a few phrases in ten years, the child learns thousands. The parrot pronounces the words in a purely mechanical way, whereas the speech of a child, even at a very early age, bears, so to speak an intelligent character [68].

Thus, the imitations by the child of what it hears give rise to auditory sensations, as well as sensations of the muscles participating in speech (muscles of the chest, larynx, tongue, lips, cheeks, etc). The exercise of these sensations, through frequent repetition becomes increasingly definite in the child's consciousness [68]. The exercise makes the previously uncoordinated muscles of speech assume definite patterns, so that the baby is able to articulate with meaning parts of the words he has heard, such as "g" for dog [148].

The faculty of listening to words, however, is but one of the conditions necessary for articulating sounds. In other words, the auditory sensations do not suffice for clear perception of words. Full perception of words requires intelligence, which is acquired mainly by the association of auditory sensations with visual and tactile ones. The more diverse the forms of this association, the more pronounced is the intelligence of speech [68]. Thus, when a child hears a sound, the reflexes of the stimulated auditory nerve include the turning of the face in the direction of the sound, and the movement of the muscles which turn the eyeball in the same direction. The first movement is the act of listening. The second movement leads to visual sensation. It is these two successive reflexes acquired by learning that constitute the elementary form of visual–auditory association. For example, a recent functional magnetic resonance imaging (fMRI) study found a region of the brain activated specifically for the cross-modality component of a combined visual–auditory task [134].

In addition, it is well known that visual sensations are easily associated with touch (tactile and pressure) sensations, that result in the notion of form and texture of objects. For example, Taylor-Clarke et al. [136] suggested that visual stimuli enhance tactile acuity through modulation of responses in the primary and secondary somatosensory cortex. Thus, when an object is presented before the eyes, the hand is brought into motion and when it encounters the external object calls forth tactile and pressure sensations. Further crossmodal links occur in selective attention directed toward visual–auditory-touch stimuli [152].

In this view, let us now suppose that the object grasped by the child is a bell. In this case along with the muscular–tactile–visual sensations caused by grasping the bell, a fourth reflex, the sound reflex is added to the three previous ones. If this process is repeated many times, the child begins to recognize the bell both by its appearance and by its sound. Subsequently, when as a result of learning, the reflexes from the ear to the tongue (speech muscle) take definite forms, and the child begins to call the bell "ding-ding." Thus, the sequence of successive reflexes, through the visual–auditory–touch and muscular sensations lead to a perfect notion of the external object [68].

We shall now consider the mosaic picture of the mother placed before the eyes of the child. The latter can see the entire picture at once. But when the optical axes of the child's eyes are directed to one particular point, for instance, to the mother's nose, the latter is seen best of all, the mouth and the eyes less so, and the feet, the farthest removed from the nose, least of all. Thus it is possible to see simultaneously the whole and the part.

Hebb [16] suggested that the brain strengthens its synapses according to how often they are stimulated. This enhances communication and information processing, resulting in learning and memory of experience. Accordingly, as a result of repeatedly seeing the mother, a visual memory center develops in the infant near the visual center, where the visual sensations derived from mother are registered as memories of learned ability [15]. Now the child recognizes the mother when she approaches and smiles. As soon as the mother notices that the child recognizes her, she introduces herself saying "mama." This is the first stimulation for language development. Subsequently, the mother shows the child how to shape its mouth and to expel the air during the pronunciation of the word Mama. The child imitates her [75].

At this early age, however, words are simply "sound objects" no different in principle to visual objects. The child learns to recognize the sound objects in the same way that he learns to recognize visual objects, through learning and memory [58]. There is sufficient implicit order in the frequency of stimuli from sound and visual objects [58] for the repeated stimulation of the auditory and visual receptive surfaces, facilitating the emergence of an electromagnetic wave, which tunnels (travels) through gap junctions connecting the auditory and visual neurons into a giant neuron (hyperneuron) making them fire in synchrony. The hyperneuron essentially unites the functions of auditory and visual neurons into a continuous cognitive process comprising an integration of visual and auditory sensations [38]. This is a reflex in the full sense, composed of sensory stimulation–cortical integration and movement of speech muscles [68].

Through this reflex arises the language center of Wernicke, in which the stimulus image for the excitation of the neurons of Broca's center is prepared. The child is now in the stage of inner, unspoken word formation. As a result of the child's repeated efforts to express the word formed within itself, connecting fibers arise and an area of the intonation center (Broca) is formed within the motor cortex for the combined activation of the oral and laryngeal muscles. The child who has recently learned to recognize the mother, now upon seeing her is learning to activate the vocal muscles through the Broca center and say "Mama." The stimulus, which the mother provides travels by way of the centers of vision, recognition, word memory and word formation, and muscle innervation [36]. Thus arises speech by the acoustic sensations, which are closely associated with the muscular sensations arising in the muscles that participate in speech, namely the muscles of the chest, throat, and especially the tongue and lips [148].

The union in the hyperneuron of the auditory and visual sensations, representing individually different cognitive states of the brain, forms the basis of conceptual cognitive states. Conceptuality then, is implicit in the relationship between the use of language in the context of recognizable objects or events to which the use refers [58].

It is noted that the union of visual and auditory cognitive states into one conceptual cognitive state is realized through frequent repetition of the same visual and auditory stimulus. This means that the capacity to reproduce sensations is based on memory storage [15, 35]. Subsequently, it seems that there is a similarity between the integration of individual sensations (cognitive states) and the formation of conditioned reflexes. The latter also depend upon memory storage through stimulation of the involved senses by the same stimulus in the same direction [68]. Thus, it appears that the formation of cognitive states of the brain and of conditioned reflexes are basically very similar or identical mental events. This means that both require the participation of the highest sensory and motor centers in the cerebral cortex for their learning and long-term memory, through frequent repetition of the reflex.

The analysis of language and speech development clearly indicates that maturation of the brain through the senses (experiences) is of prime importance. Maturation is related to intelligence, which includes language understanding, reasoning, perception, problem solving and learning ability [149]. Early learning of intricate movements of the lips, tongue and cheeks, and in their complex coordination with swallowing and breathing rhythms, resulting in a mobile mouth, is of prime importance for the development of speech [148].

Specifically, when we speak we coordinate the movements of the lips, tongue and the shape of the mouth with the sounds we hear. It has been suggested that this synchrony is achieved as a consequence of traveling waves or solitons uniting the neural activity in different parts of the brain. At the molecular level, solitons are quantum coherent structures—a unified state of energy/matter. Biological processes represent a unique synthesis between energy and structure such that there is no discontinuity between them. Solitons or traveling waves are involved in muscle contraction, protein folding, DNA "zipping" and "unzipping," action potentials of nerves and, crucially, in the brain itself. It has been suggested that the brain is an excitable medium, which is structured by the senses, and that cognition (and all mental processes) correlates with the spatiotemporal evolution of traveling waves in the brain [58].

The higher levels of cognition, such as language, can be understood as an essentially inductive process, involving a succession of steps (neural events) that mediate implicit relationships between established cognitive processes. Once the lower levels of cognition have been established, the implicit relationships between these states of neurocognition may become explicit at higher levels of cognition, such as language, conceptuality, etc, which underlie biological structures, for instance the Wernicke area of inner speech, that combine metabolism–function and structure–energy [58].

This view, that the higher cognitive states of the brain develop through maturation of successive levels of cognition, each level being characterized by structural and functional transformations of the cerebral cortex [58], could also be applied to the peripheral sensory receptors, since the cerebral cortex is regarded as having been built by the neural activity of the cerebral terminals of sensory receptors [35]. The sensory receptors of a newborn child might undergo successive stages of maturation through learning of experience, which might be correlated with the development of successive levels (states) of cognition of the brain. In support of this view may be that at the present time only a few aspects of the codes employed by the nervous system to transmit information have been identified and evaluated. For example, it is known that each sensory modality stimulus is a composite of several components. Each of the components evokes several codes, which are transmitted to the processing centers [36].

In sum, it is pointed out that language and speech development is structured as a consequence of auditory, visual, tactile, and muscle spindle stimuli [214] and their crossmodal interaction [15, 38]. The corresponding chemical and structural changes of the brain, depending on the stimulus, comprise successive cognitive changes or states of the brain, each state mediating the state immediately above [58]. The abilities to acquire language and speech correspond to highest levels of conscious cognition and motor control, respectively. There are several theories as to how conscious cognition might arise. The most accepted theory of consciousness or perception is that of Penrose and Hameroff [38, 113]. The authors suggest that multisensory experiences are integrated and reach a "threshold" for perception or consciousness in the hyperneuron. The hyperneuron is a giant neuron comprising neurons from many areas of the cerebral cortex connected by gap-junctions (electrical synapses), in which the component neurons fire in synchrony. McFadden [85], however, suggests that the massive discharge of cortical neurons through the input of multisensory experiences might generate a conscious electromagnetic field in the brain, which initiates conscious movements of the skeletal muscles, including speech muscles.

Mastication in Man

6

6.1 The Central Pattern Generator

Current authors view mastication as a cyclical movement, like breathing or walking. The basic rhythm of cyclical movements is driven by a program that is hard-wired in the "central pattern generator" which is located in the reticular formation of the midbrain at a site strategically placed near the trigeminal nuclei and the nuclei reticularis pontis oralis and caudalis [78, 172, 219]. The central pattern generator is probably programmed by the conscious electromagnetic field generated by the synchronous firing of cortical neurons (see Sect. 4.8).

Like voluntary movements, cyclical movements are normally fine-tuned by sensory signals from the masticatory muscles and oral receptors acting through the action of reflexes. There is very good experimental evidence that cyclical masticatory movements are neither purely voluntary nor purely reflex. Mastication manifests itself as a patterned cycle of muscle contractions, which in turn produce the characteristic three-dimensional envelope of jaw movement, coordinated tongue and cheek movement, and compressive and shearing forces between the teeth to comminute the food bolus making the act of chewing as we know it [78, 172].

It should be emphasized, however, that the central pattern generator is principally a timing mechanism for creating rhythmicity coupling the movement of the jaw and tongue. As such it does not produce mastication or chewing in the true sense, in which the development of specific paths of movement, the altered timing sequences, and the interocclusal forces are characteristic additions to rhythmicity per se [78, 218]. During normal rhythmical chewing movements all the jaw-closing muscles on both sides are activated at about the same time through the coupling of the hyperneurons in the left and right hemispheres (see Sect. 4.2). During the opening movement only the jaw-opening muscles are active. When chewing occurs on the right side of the mouth the activity of the left masseter during the chewing stroke is less than the activity in the right masseter [78].

In sum, the important components of the central pattern generator include the motor cortex, which activates the premotor neurons, which in turn activate the trigeminal motor nucleus, which cause the rhythmical chewing movements. The rhythmical chewing movements generate sensory signals through activation of the oral sensory receptors the input of which activates the premotor neurons following the sensorimotor integration function in

M. Z. Pimenidis, *The Neurobiology of Orthodontics*,
DOI: 10.1007/978-3-642-00396-7_6, © Springer Verlag Berlin Heidelberg 2009

the somesthetic cerebral cortex. Thus, the activity of the central pattern generator can be triggered both from the cortex and from sensory signals from the masticatory system. It is noted, however, that feedback signals from the masticatory apparatus are not necessary for continued activity of the central pattern generator. In other words the chewing center can continue to send out its signals to the motor nerves of the muscles even when the rhythmical chewing movements of the muscles are pharmacologically prevented [78, 218].

For example, in this particular experiment, rabbits were not only given a general anesthetic but were also paralyzed with a curare-like drug that blocked the neuromuscular transmission of action potentials to the masticatory muscles without, however, interfering with the action potentials that traveled out from the brain along the motor nerves. The curare-like drug stopped the action potentials from activating the chewing muscles. Hence, the drug prevented the jaw from moving in the normal rhythmical masticatory pattern when the animal's brain was stimulated. It was concluded that the central pattern generator does not depend on sensory feedback to continue the rhythmical chewing movements [78, 219].

The central pattern generator, however, is unable by itself to adjust the muscle force required to break down food whose texture may be unpredictable and changes from one chewing stroke to another. For this reason the masticatory system is powerfully modulated by reflexes, which automatically fine-tune the centrally generated masticatory movements in order to control the force of the muscle contraction. The most important of these reflexes is the stretch reflex and the reflexes emanating from the periodontal mechanoreceptors, which are described below and in Chap. 8, respectively. It is interesting to note that reflexes emanating from the mechanoreceptors of the skin may also modulate masticatory muscle function. For example, recordings from different nerves innervating the mechanoreceptors of the skin of the face, lips and oral mucosa stimulated by voluntary movements of the jaw through deformation of the tissues have indicated that there are tactile receptors activated only by jaw-closing movements, suggesting that proprioceptive signals from the face and mouth may specifically control the opening and closing rhythm of the mouth during chewing [23, 78]. The reader is reminded that the jaw-opening muscles as well as the facial muscles have no muscle spindles and their function may be substituted by the mechanoreceptors of the face and oral mucosa (see Sect. 8.12).

It is also interesting to note that a number of authors have suggested that the function of mastication does not gradually develop from the suckling function, although the phasic alternation of cyclical masticatory movements are reminiscent of the cyclic function of the act of suckling. Rather, the development of the mature patterns of biting and chewing appear as new learned functions that may be triggered by eruption of the teeth, reflecting coincident maturation in oral sensory experience and in the progressions of encephalization [1, 65] (see Chap. 10).

6.2 The Trigeminal Stretch Reflex

The stretch reflex in the trigeminal system functions to adjust the force exerted by the jaw-closing muscles to compensate for changing resistance to closing that occurs during chewing. It also plays a role in maintaining the posture of the jaw in the "rest position" during vigorous head movements, for instance during running. The stretch reflex, like other somatic reflexes in the body, consists of a sensory receptor, an afferent or sensory pathway, an integrating center in the central nervous system, where the sensory neurons synapse with the motor neurons, and an output pathway consisting of the axon channels of pyramidal motor neurons, which innervate the masticatory muscles through the motor nuclei of the trigeminal nerve. The receptors for the stretch reflexes are muscle spindles, which are located in the muscles in parallel with the muscle fibers. The jaw-closing muscles, but not the jaw-opening muscles, are richly endowed with muscle spindles [78, 217].

The sensory nerves that carry the signal from the muscle spindles to the brain are the largest myelinated nerve fibers found anywhere in the body, the so-called group Ia afferent neurons. The trigeminal Ia afferents pass into the brainstem via an anatomically unique path. Although they are sensory fibers, they enter the brain in the same nerve trunk as the motor nerves that are leaving the brain, rather than with other sensory neurons as is the case in the spinal cord. Their cell bodies lie within the brain, in the mesencephalic trigeminal nucleus, which in turn sends fibers to the trigeminal motor nuclei where they form excitatory synapses with the alpha motor neurons that innervate the jaw-closing muscles [78, 217].

Because both the nerve fibers that carry the action potentials from the muscle spindles to the mesencephalic nucleus of brainstem, and the fibers that bring action potentials from the brainstem back out to the masticatory muscles are large and myelinated, they conduct action potentials at very high velocities (about 50 m s^{-1}). Also, because the distances over which action potentials must travel in this reflex are quite short (about 90 mm in each direction), the stretch reflex occurs very quickly (about 7–8 ms). The speed of the stretch reflex response enables the muscles to react extremely quickly to the changes in the resistance of the food being chewed. The chin-tap reflex is used by neurologists to test the normal function of the sensory and motor nerves in the trigeminal system [78].

During any normal skeletal muscle contraction, the brain also sends signals along the gamma motor neurons, which innervate the ends of the muscle spindles causing them to contract too, through their own actin-myosin contractile system. The contraction of both ends of the spindle stretches and hence activates the middle part of the spindle, where the sensory receptor (the primary or annulospiral ending) is located, causing action potential signals to be sent along the large myelinated Ia sensory fibers to the mesencephalic nucleus of the trigeminal nerve. Then the action potentials reach the jaw-closing muscles causing them to contract (to twitch), through the motor nuclei of the trigeminal nerve [78, 217].

It should be noted that the muscle spindles also have “secondary” sensory endings, which are anatomically different from the primary endings. The primary endings are sensitive to changes in muscle length and are rapidly adapting receptors. The secondary endings give a conscious signal related to muscle length and, therefore, are slowly adapting receptors [78].

6.3 Isometric Contraction of Jaw-Closing Muscles

The isometric tension developed by a muscle during maximal voluntary effort corresponds closely to that produced by tetanic stimulation of its motor nerve. Therefore, activity of the elevator jaw muscles during maximal voluntary clench of teeth may indicate their maximal strength [125].

The strength of voluntary contraction is increased both by an increase in the rate of individual motor unit discharges and also by recruitment of more motor units. Since each action potential represents activation of a motor unit, a correlation can be expected between muscle force and its pattern of electrical activity [125]. In order to understand how isometric contraction controls biting force, consider what happens when you bite down onto brittle food such as a raw carrot.

The brain sends signals along the alpha trigeminal motoneurons to the jaw-closing muscles. This causes the muscle to contract and hence, to shorten and move the lower teeth towards the food. At the same time, the brain is also sending signals along the gamma motor nerves innervating the muscle spindles in the jaw-closing muscles. These cause both ends of the spindle to contract, so that each spindle shortens at the same rate as the skeletal muscle fibers, keeping the spindle under steady tension. When the teeth touch the food, the resistance to closing movement suddenly increases and the muscles stop shortening. However, the muscle fibers are still actively contracting, although they are prevented from shortening. The contraction of the muscles now becomes isometric, meaning that they are contracting under constant length. When, however, the resistance of the food during chewing increases, the gamma motor neurons of the brain continue to send signals that further shorten the spindle and hence further stretch the elastic center where the sensory receptor lies. This increases the activation of the receptor which sends action potentials to the mesencephalic nucleus, which in turn are projected to the trigeminal motor nucleus causing the jaw-closing muscles to contract more strongly, resulting in an additional biting force on the food.

All this happens in a small fraction of a second and automatically, in other words without any conscious intervention. Thus, the resistance of the food being chewed informs the brain through the muscle spindle receptors of the masticatory muscles that the food is tough, and the brain rapidly learns about the texture of food and automatically resets the reflex response accordingly [78, 219].

Numerous studies have demonstrated a rectilinear relationship between electrical and mechanical activity during isometric contraction. This implies that the electrical and mechanical activity recorded during maximal clenching of the teeth are alternative measures of the strength of the jaw-closing muscles. However, the question of whether the maximal bite force in fact represents the maximal strength of the jaw-closing muscles (i.e., the output derived from the motor units available working at their maximal rate of discharge) may be raised [125]. Rasmussen and Moller [185] tested the maximal clenching of the teeth in the intercuspal position by determining the electrical activity in the anterior temporal and masseter muscles, and found that the maximal bite force does not represent the maximal strength of these muscles. They then suggested that there exists for each individual a maximal level of attainable muscular strength. Below this level the actual strength varies with the degree of physical activity. The effect of training on isometric endurance and maximal voluntary contraction supports this suggestion. Isometric endurance is related to fiber type composition both before and after training, but the effect of training is not. Therefore training does not level out differences between individuals with respect to endurance.

Functional disorders of the chewing apparatus are characterized by tenderness and pain of the muscles of mastication. The endurance and strength of the masticatory muscles may determine susceptibility to tenderness and pain. For example, nonfunctional activities such as bruxism (clenching and grinding of the teeth) may be responsible for tenderness and pain in the muscles of mastication. Indeed the comparison of the strength of the temporal and masseter muscles between subjects with and without such symptoms has shown that patients with functional disorders of the masticatory muscles have significantly lower values of maximal bite force compared to normal subjects. The affected and unaffected sides do not differ in patients with unilateral localization of the disorder. Reduction of signs and symptoms due to treatment is followed by an increase in maximal bite force [125].

6.4 The Unloaded Reflex

A dramatic example of biting on a very hard brittle object, for instance a nut shell, is the unloaded reflex. In this instance, the brain is sending ever increasing signals to the trigeminal alpha motor neurons that innervate the jaw-closing muscles, as well as to the contractile part of the muscle spindles through the gamma motor fibers, resulting in harder contraction of muscles. When the nut shell cracks the resistance of the jaw-closing muscles suddenly falls to a low level or disappears altogether. At this point, the jaw-closing muscles are still active and contracting hard, and there is the potential for the teeth to cause a lot of damage to oral structures. However, this does not happen.

First, because when the nut shell cracks the jaw begins to close as the muscles, now unresisted, begin to shorten at high speed. This shortening of the jaw-closing muscles removes the tension from the middle of the spindle and hence, the primary sensory receptor stops sending action potentials through the Ia afferents to the trigeminal mesencephalic nucleus, and the muscles are deactivated. This is the unloaded reflex, which protects the oral structures from damage. The second reason is, that when the jaw-closing muscles bite on the nut shell they are making powerful isometric contractions. This activates the jaw-opening muscles. Then, when the nut shell cracks and the jaw begins to close unresisted, the tension in the tonically active jaw-opening muscles keeps the mandible from springing too far upwards [78].

6.5 Voluntary Control of Masticatory Muscles

The main focus of the discussion so far has been on the control of the masticatory muscles by the central pattern generator through central motor commands, and by the various peripheral reflexes that fine-tune the output of the central pattern generator. The masticatory muscles, however, like most other skeletal muscles are also subject to highly precise voluntary control, following patterned oral experiential sensory input to the cerebral cortex [85, 189]. This implies that the neural development of the mouth is concomitant with the process of maturation of the decoding of information and learning mechanisms in the cerebral cortex [43, 67].

In the voluntary movement plan the output of the motor cortex bypasses the central pattern generator to reach directly the trigeminal motor neurons, which send action potential signals to the masticatory muscles. Then voluntary masticatory movements and forces can be generated without invoking rhythmicity. This is essentially the process that controls other skeletal muscles as well, with an important difference. While it is common to move, for instance, one finger on one hand, the masticatory muscles of both sides are activated simultaneously during almost every voluntary jaw movement through coupling of the left and right hyperneurons (see Sect. 4.2). We can voluntarily start chewing movements of the mouth, yet in most circumstances of normal mastication a conscious or voluntary effort is not required to maintain the masticatory activity [76].

6.6 The Closed-Loop Theory of Motor Control

A prevailing problem in motor coordination is how the motor systems are organized and regulated. In this context, there are two antithetical mechanisms that adjust the muscle force required during chewing. One of the foremost views, and until recently the favored position, is that peripheral receptors are necessary and sufficient for motor control through learning. This "closed-loop" theory proposes that sensory feedback, error detection and error correction are necessary requirements for movement regulation. The motor command that specifies the response to be generated is executed. Feedback through the muscle spindle and joint receptors checks for discrepancies between the output specified and the movement actually produced (error detection). The errors are evaluated and the central nervous system attempts to minimize the size of the error (error correction). Typically, slower movement sequences, for instance slow movements involved in speech learning to pronounce words as well as other sounds, are under closed-loop control [186].

The closed-loop control theory also suggests that feedback augmentation i.e., an increase in the amount of the sensory input, will lead to an increase in motor performance. In other words, the motor task is performed more accurately than without sensory augmentation. Such increased sensory input occurs during the period of early mixed dentition when the roots of the permanent incisor teeth are incompletely developed, as suggested by Greenberger [45], who found that the tactile threshold for young incisors is lower than the threshold for incisors with completely formed roots. These findings would support the contention that occlusal feedback in the mixed dentition plays a more important role in mandibular movements because of the less favorable crown/root ratio of incisor teeth. In this view, early treatment of certain malocclusions of the teeth, for instance of the anterior crossbite of teeth, may give more favorable results than at later ages.

The early sensitivity of the incisor teeth with the augmented sensory input, however, does not seem to be compatible to the clinical observation that most young children when asked to close the teeth bring the lower incisors into an edge to edge relationship with the upper incisors, or even into an anterior crossbite position, suggesting that young children, despite the augmented incisor feedback, are unable to use the incisor reference to guide their mandible accurately into the intercuspal position. This may imply that the sensory input from the anterior teeth does not contribute substantially to guiding the occlusion of the teeth, at least in the mixed dentition period, since the oral behavior of most children is clumsy, as the specialist clinical observation clearly suggests. Specifically, in the period of mixed dentition the mandibular movements are slow and exploratory, and the chewing movements are uncoordinated, suggesting learning procedures in progress.

Accordingly, it has been suggested that a human is not born with correlated information of muscular activity in skilled chewing movements after a particular efferent pattern is sent downstream. On the contrary multiple levels of cognition must gradually build up through multisensory oral, visual, and vestibular experiences, as well as through the augmented occlusal feedback of the early dentition (see Chaps. 4 and 5). If this view is correct we may then expect slow learning procedures lasting until the permanent dentition is completed through the gradual maturation of the functions of the cerebral cortex. During this period one should not, however, exclude the possibility that the greater sensibility of the incisor teeth in the mixed dentition may become a predisposing factor for an abnormal relationship of the teeth, as for instance, in the anterior cross bite, through an abnormal habitual closure of the incisor teeth coupled with or without a favorable skeletal pattern.

It has to be noted also that, since the maturation of the skilled chewing movements is a long process, it is conceivable to agree with the suggestion of many early investigators that in early postnatal age the behavior of the muscles is under the influence of the primary innate motor patterns, which have not yet been overridden by the secondary motor patterns learned by individual design [187].

6.7 The Open-Loop Theory of Motor Control

The closed-loop motor control theory has been challenged on the grounds that the motor acts are centrally regulated and may occur in the absence of feedback. The open-loop theory proposes that information from peripheral sources enters the higher centers of the central nervous system where it is processed and transformed into information necessary for patterned movement. The higher centers dictate the characteristics of the desired movement, for example the exact motor commands for each muscle group and their exact temporal sequence. In the open-loop view, the efferent command (motor program) is all that is necessary to control coordinated movement. The implication is that peripheral feedback is unnecessary for the regulation of movement. Typically fast movements, for instance the skilled movement of chewing or speaking, are under open-loop control [186].

However, the closed-loop mechanism is more accurate but slower; it requires more time to function than the open-loop predictive mechanism of the central nervous system. The central pattern generator is programmed at least for slow and fast chewing in different subjects. During slow chewing the open-loop control mechanism predicts the muscle force required during a chewing cycle. This mechanism, however, explains only about 15% of the force adjustment to the toughness of a bolus of food, while the feedback closed-loop explains most (85%) of the force accommodation [78].

6.8 Orthodontic Implication of Motor Control Theories

There are substantial amounts of evidence for both the closed-loop and open-loop mechanisms. Rapid adjustments to steady-state oral functions occur as a result of preadjustments of the descending multimotor commands. This view, which supports the open-loop control, suggests the existence of some internal monitor of the motor command for rapid error correction, independent of sensory feedback. On the other hand, compensatory responses to unanticipated responses require the use of afferent feedback. Thus, the role of the oral sensory feedback lies in its ability to effectively modify the motor output through error detection and correction, and hence it is used in the restoration and relearning of movements [186]. For this reason when an individual is engaged in a therapeutic procedure, namely an orthodontic therapy for a malocclusion of the teeth associated with motor dysfunction, the movement of the jaw after correction of the occlusion of the teeth is slow, thereby permitting ongoing feedback control, enhancing error detection. Once errors have been detected the output (motor) plan can be modified in an attempt to correct the errors and achieve the desired movement. Once such exercises have led to mastery of proper movement the motor control may be relegated to automatic control or open-loop control of movement. The ever-changing movements within the oral structures require that the central processes constantly be informed of the actions of each oral structure. This may be accomplished by "ongoing" closed-loop control or may be a function of the motor command itself.

In this context, in early age malocclusions of the teeth with premature contacts, in an ever-changing occlusal pattern, the movements of the mandible are slow because they depend upon adequate sensory feedback for error detection and correction of movements. In such circumstances the orthodontic therapy helps to correct the occlusal pattern and hence allow the normal maturation of skilled oral functions of biting, chewing, and speaking, involving centrally coordinated muscle function. Thus, early orthodontic treatment may be essential in order to restore the actual pattern of neural innervation that underlies the learning of oral coordinated chewing movements.

6.9 Multicontrol Levels of Coordinated Chewing Movements

The programming of coordinated movements may not be in terms of individual muscles, as classically believed. This is because any model of such hierarchical organization of motor control, in which an executive center specifies the action plans for selected muscles, would be extremely cumbersome considering the large number of degrees of freedom available, for instance, for the mandible. The problem is exaggerated even more by the variability of the innervation pattern of muscles and joints [186].

An alternative to the hierarchical model is the "heterarchy" in which, there is more than one center for control of coordinated motor activity. These various centers can act independently, in other words they are autonomous subsystems which rely not on coordinated movements but on coordinated structures, generally referred to as a group of muscles, often spanning several joints, which act as a unit, and allow multilevel response organization. In this type of control system each group of muscles solves a limited aspect of the degrees of freedom problem without direct specification from the brain. Such multicontrol levels of the oral structures are most evident in speech production, in which the different output parameters involved in the generation of speech are not programmed at the same level in the central nervous system [186].

6.10 Central Sites Eliciting Mastication

The cyclic chewing movements of the jaw and tongue can be elicited by electrical stimulation of the basal ganglia, corticobulbar tracts, and parts of the limbic system, but are only ultimately dependent upon the integrity of the brainstem reticular formation. This implies that much of the information for mastication is generated in suprabulbar areas, which are funneled to the trigeminal motor neurons through the central pattern generator [76, 172].

The role of the cerebral cortex, however, for the initiation of chewing movements of the jaw is uncertain. Experimental studies in human subjects have indicated that masticatory movements are not located in the sensorimotor cortex in the precentral and postcentral gyri of the cerebral cortex, as are the nonmasticatory movements of the mandible, but seem to have a separate representation [66]. This may imply that the cyclic chewing movements are not perceived in the cerebral cortex as are the voluntary chewing movements. This may also be the reason that we bite the tongue or cheek accidentally during chewing if our attention is distracted.

The role of the cerebral cortex in mastication is but slowly being explored. It has been proposed that the cerebral cortex is involved in initiating the jaw-opening muscles, elaborating, integrating and perhaps engramming peripheral sensory information, regulating the generating of interocclusal force and the intimate linking of this kind of activity with that of the central pattern generator of the brainstem. In addition, it has been suggested that maturational and learning processes may involve cortical activity [172].

The unclear role of a variety of other central sites upon the generation and regulation of the masticatory cycle is a limiting factor in our understanding of the act of chewing. For example, Dellow [188] has referred to satiety and feeding centers in the ventromedial and lateral hypothalamus, respectively, as links to the limbic system and mesencephalic structures. He has also pointed out that the amygdala, septum, hippocampus, ventral tegmentum and central grey matter are all involved in the control of mastication.

6.11 Tonic Stretch Reflex and Postural Position

The term tonic is used to indicate that the stretch reflex is continuously active, as one would anticipate for a system involved in limb postural muscle contraction. An adequate stimulus for the tonic stretch reflex may be looked upon as the organism's opposition to gravitational pull. In opposing gravity certain muscles are maintained in a stretched state and at relatively constant elongations. These natural muscular stretches result in deformation and hence activation of the muscle spindle primary endings. These sensory receptors send action potentials along the group Ia sensory nerves to their synapses which excite (depolarize) the alpha motoneurons which innervate the muscles causing them to contract [214].

It has been argued that the tonic stretch reflex operates continuously at a low level in the masticatory muscles to keep the mandible in its so-called "rest position" (or postural position) relative to the maxilla even when the head is stationary. This is an important issue because the rest position is a reference position in orthodontics and prosthodontics. Despite the attractiveness of this notion, recent research has shown that the vertical position of the mandible is maintained passively in its postural position by the elastic forces from the perioral soft tissues (which includes those in the muscles). It is actually more accurate to call these forces "viscoelastic," as the viscosity of the perioral soft tissues dampens down the springiness that would result from purely elastic recoil. These passive forces are sufficient to maintain the mandible in or near its rest position even when the head moves up and down during waking, but by stretch reflexes when the head is moving more vigorously up and down during running and jumping [78].

6.12 Control of Tonic Stretch Reflex

Voluntary movements are thought to be organized and initiated in motor areas of the cerebral cortex. The motor cortex, like the cerebellum, exerts a detailed control function over the spindle fusimotor system as well as the alpha motor system. Thus, the motor output emanating from the cerebral cortex is codistributed between the alpha and gamma motor systems in such a manner that fusimotor activity cooperates functionally with alpha activity. With respect to alpha and gamma cooperation in cortically initiated and sustained voluntary movement, two divergent theories have been developed.

The "follow-up length servo theory" suggests that voluntary movement is initiated indirectly via the fusimotor system. Specifically, the theory suggests that the motor output descending from the cerebral cortex initially activates the gamma motor system, which controls the amount of spindle intrafusal fiber shortening and therefore the change in length of the extrafusal muscle fibers. (It is noted that fusimotor fiber activity has been shown to be equivalent to a certain length change in the whole muscle.) Given this equivalency, by appropriately grading the central drive to fusimotor (gamma) neurons, the motor cortex can control the amount of spindle intrafusal fiber shortening and therefore the change in length of the extrafusal muscle fibers. The desired movement then is represented in the gamma neuron population as a demand for a specific length change. The discharge of the spindle primary endings produced by the intrafusal fiber shortening then reflexly drives, through the monosynaptic arc, the alpha motoneurons and extrafusal fibers to develop sufficient tension to satisfy the length demand of muscle fibers. Thus, the tension-generating alpha system follows the length demand impressed upon it by the initial fusimotor neuron discharge [214].

In this view the tonic stretch reflex in maintaining a static or postural limb position describes a servo system. In essence a servo system is an automatic closed-loop comparison system, in which the desired position of a limb or the command movement of an object is compared with its actual position or movement, and action is automatically taken to correct any difference between the desired and actual states. The stretch reflex acts to maintain a proper balance between input from the spindle and alpha motor output so that the contractile force developed in muscle just counterbalances the pull of gravity on the limb. In maintaining a fixed posture, however, transient muscular imbalances tend to deviate the limb from the desired position. Any length change in the muscles participating in maintaining the posture will be detected by the spindles as error in the servo loop. The difference in the frequency of firing of the primary ending when the muscle is at the desired length and when it has been stretched comprises the error signal, which is passed monosynaptically to the alpha motoneurons resulting in contraction (shortening) of the same muscle. The muscle will continue to shorten under the action of the stretch reflex until it is again returned to its equilibrium, or null length. Thus, the closed-loop control cycle provides a negative feedback wherein the muscular contraction (monitored as length) opposes the unwanted length change [214, 215].

6.13 Cerebral Cortex Control of Voluntary Movements

In voluntary movements the spindle primary ending responds not only to a maintained stretch (tonic reflex) which opposes the gravitational pull, but also to the rate at which the muscle lengthening takes place. So far, nothing has been said regarding the regulation in the stretch system when a rhythmic rapidly alternating pull and release (i.e., a sinusoidal stretch) is being applied continuously to a muscle. Servo systems inherently suffer from transmission delays around their control loops, most notably those involved in transmitting impulses to and from the central nervous system and in the time required for muscle to develop full contractile tension following the arrival of motor impulses at the myoneural junctions. Thus, there are several objections to the concept that the fusimotor system could initiate voluntary movements on a follow-up length servo basis.

Accordingly, it has been suggested that voluntary movements can be initiated via those fibers descending directly from the cortex that monosynaptically activate the alpha motoneurons. This implies that cortical control loops are of major importance in skilled voluntary movements. The cortical motor control is of course consistent with the general process of encephalization of movement in humans as well as with the experimental observations derived from primate research. In the case of hand movements, instructions issued from the cerebral motor cortex are applied directly to the alpha motor neuron system via the monosynaptic corticomotoneuronal projection (pyramidal pathway) resulting in muscular contraction. In turn, sensory information from muscle spindles reaching the cerebral cortex concerning discrepancies between intended and actual muscular displacement results in discharge of specific cortical motor cells. This discharge is then transferred back to the alpha motoneurons with maximal speed, again via the pyramidal system.

Thus, both in the initiation and maintenance of voluntary movement, cortical control loops are dominant. In this scheme, the function of cortical regulation of the fusimotor spindle system (efferent) is to maintain the flow of spindle information back to the motor cortex and cerebellum, as well as to other central motor structures. This is achieved by cortical coactivation of the fusimotor neurons, subserving spindles in the same muscles to which the alpha command is issued. While the fusimotor system thus does not appear to be involved in the initiation of voluntary movement, it does appear to provide the necessary background for the spindle afferent regulation of such movement. In other words, cortical control over the fusimotor system functions to maintain the flow of spindle information to, in particular, the motor cortex, by ensuring that the spindles do not stop discharging when the muscle contracts [212, 213].

In sum, any muscular movement is an expression of the integrated action of the alpha and gamma motor systems. The gamma innervation of the spindle places the discharge of spindle afferent endings under the regulatory influence of the central nervous system. The gamma control is independent of the alpha control. The gamma control is capable of selectively regulating the velocity and length responses of the spindle afferent endings as well as controlling their response sensitivity. Functional cooperation between the two motor systems is a precondition for smooth, efficient motor performance. The motor centers involved in the initiation, integration, and maintenance of muscular activity must, therefore, have some means for controlling the activity of both systems simultaneously. Such parallel regulation is accomplished in large part by the two suprasegmental motor structures, the cerebellum and the cerebral cortex [214].

6.14 Cerebellar Control of Voluntary Movements

That the cerebellum is intimately involved in the integration of motor acts is clear from the striking alterations in posture and in skilled voluntary movements characteristic of intentional ataxia, dysmetria and incoordination following cerebellar lesions, and its pathways [213].

The cerebellar cortex receives inputs pertaining to muscle length and tension as well as to joint position. In addition there is a large input from cutaneous receptors to the cerebellum. The same areas of cerebellar cortex are in turn, reciprocally connected to the cerebral motor cortex so that not only does the latter receive information about cerebellar activity but, equally significantly, the cerebellum is informed about present, and perhaps ensuing, motor commands being issued from the cerebral motor cortex concerning motor behavior [213].

In addition, evidence indicates that the cerebellum is largely responsible for maintaining the organization of the alpha–gamma linkage that ensures the integrated action of the two motor systems. Functional cooperation between the alpha and gamma systems is destroyed by cerebellar lesions. Such injury typically is associated with severe depression in the level of fusimotor (efferent) output due to the elimination of indirect cerebellar facilitation of the fusimotor system. In the absence of cerebellar fusimotor facilitation, the stretch reflex is depressed, and the muscles become flaccid or hypotonic, offering virtually no resistance to stretch. In contrast, lesions in other central structures result in an abnormally high level of fusimotor output leading to hypertonic stretch reflexes. Injury to the basal ganglia, to specific areas of the cerebral motor cortex, or to certain of the pathways descending from the motor cortex, all lead to such hypertonic stretch reflexes. Lesions at the latter two sites typically produce a spastic syndrome characterized by hypertonicity in the antigravity muscles, or physiological extensors; basal ganglia lesions generally are associated with Parkinsonian rigidity in which the muscular hypertonicity and exaggerated stretch reflexes are present in flexor and extensor muscles alike [214].

That the limb muscle spindle is indeed a focal structure in the cerebellar regulation of normal motor behavior is evidenced by the fact that recovery from the gross motor deficiencies consequent upon cerebellar lesions is correlated with a return of muscle spindle responses, especially of gamma motor activity. With the recovery of fusimotor bias, alpha activity can again be directed by spindle input, and the temporal relationships between muscle length changes and alpha motor output reapproximated [212, 213].

In contrast, lingual spindle primary afferents apparently do not project to the somatotopically organized cerebellar cortex as indicated by studies with recording electrodes in the monkey. The finding that the spindle afferent fibers of the tongue muscles do not have a cerebellar representation indicates that the lingual spindle projection is differentially organized relative to the limb spindle projection in terms of its central termination sites. Thus, aside from reflexive control, tongue movement appears to depend most intimately on the highest motor control center in the brain, the cerebral cortex. This conclusion is consistent with the progressive encephalization of those movements subject to increasing degrees of voluntary control, and indicates that the process involves not only an increasing extent of cortical control but also the establishment of qualitatively unique control circuitry. One is then led to the further conclusion that movement encephalization is observable not only in ascending the phylogenetic scale but also within different muscular systems of a single species [214].

6.15 Conscious Sensation of the Muscle Spindle

Until recently it was considered that the existence of a cortical projection for a given afferent system constitutes evidence that the system participates in conscious sensation and discrimination. This notion is no longer entirely valid. The argument that certain authors have presented is that the muscle spindle is not involved to any significant degree in the arousal of sensations of position and movement in the cerebral cortex, this function being subserved by the capsular and pericapsular end organs situated in and around the joints. Accordingly it has been suggested that the primary function of the muscle spindle is in the automatic, or unconscious, regulation of postural reflexes as well as in phasic stretch reflexes such as the tendon jerk reflex [214, 216].

The idea that limb spindles are not responsible for evoking sensations of position and movement raises the question of how we are to classify spindle relative to joint receptors. The spindle, of course, is traditionally categorized as a proprioceptive end organ along with the capsular and pericapsular receptors; yet limb spindles do not contribute to the arousal of a proprioceptive sensation which, as classically defined, is position–movement sensibility. Historically, position–movement sensibility has been described under a variety of terms, the more favored being proprioception and kinesthesia. These terms should then define capsular inputs of joints, while the spindle function defines primarily the unconscious (spindle) regulation of limb movement [214].

The separation of afferent input into a conscious–unconscious dichotomy, however, tends to compartmentalize our thinking rather than call attention to the fact that the cerebral cortex functions as an integrated sensorimotor unit. It is extremely important to realize that all afferent information arising in response to or associated with movement is utilized by the voluntary motor system in its integration and control of muscular contraction. In the primate, this is true for spindle, capsular and cutaneous input, as well as for visual input in the case of the extremities and for auditory input in the case of the speech musculature. It is equally important to realize that the question of whether this sensory information also impinges on perception or consciousness is an entirely different matter. This means that activity in certain ascending neural pathways engages a specific set of cortical neurons (neural assembly) into synchronous firing exceeding all other neural assembly activities in some type of competition, which in turn meets the cost of entering into perception or consciousness. In other words the particular neuron assembly becomes conscious [38] (see Sect. 4.2).

This is a somewhat broader interpretation in that it does not base movement organization exclusively on the afferent systems and cortical regions engaged in conscious perception but also includes those inputs, ascending relays, and cortical regions not involved in perception per se. This idea avoids the implication that all movement organization is a consciously directed process and, therefore, may take advantage of more input modalities, especially spindle input, which apparently does not impinges on perception or consciousness [214].

6.16 The Subconscious Nature of Complex Movements

In the voluntary act of mastication the cortical mechanisms are concerned with the initiation and halting of masticatory movements, as actually occurs for all voluntary movements of the body. However, the cortically initiated movements of the muscles of the body are crude and irregular if there is damage to some other region of the brain, especially the cerebellum. For example, subjects with a cerebellar lesion complain that while the movements, for instance, of their left hand on the normal cerebellar side, are done subconsciously, they have to think out each movement of their right arm on the side of the lesion [37]. Thus, normally most complex movements of the body are carried out subconsciously with consummate skill. The more subconscious we are of the actual muscle contractions concerned in complex muscle movements, for instance a golf stroke, the better it is, and the same with tennis, skiing, skating, or any other skill. In all these performances we do not have any appreciation of the complexity of muscle contractions and joint movements. All that we are conscious of is a general command, such as "place finger on nose" or "write a signature" and the whole performance flows automatically from that. The cerebellum is concerned in all this enormously complex organization and control of skeletal muscle movements, and perhaps of the masticatory movements too. Throughout life, particularly, in the early years, we are engaged in an incessant teaching program for the cerebellum in order to carry out the skilled movements required for all of a remarkable range of tasks [37].

The function of the trigeminal sensory and motor systems are very analogous to those of the hand. For example, the control mechanisms for the precise manipulation of small objects with the teeth appear analogous to those used for precision manipulation of small objects by the thumb and fingers. This is reflected in the similarity of the tactile acuity and tactile sensitivity of the oral and hand tissues (see Chap. 8). The information contained in the responses of the mechanoreceptors in the periodontal ligament trigger the release of a central command that is used for motor control. All that we are conscious of is a general command "chew: and the whole performance goes automatically from that" [78, 93].

6.17 Masticatory Plasticity

The masticatory system described so far could hardly be described as a rigidly structured mechanism for comminuting food. Its behavior is flexible enough to accommodate a range of external (peripheral) and internal (central nervous system) perturbations whether these be transient or permanent; and the observations of a particular pattern adopted by any subject presumably reflects the sum of all such influences. It is this fact that makes interpretation so difficult.

Events influencing the final form of the chewing cycle presumably fall into two broad categories: those benign influences which occur naturally or fortuitously in the course of time, and less natural ones which are later injected by fate or by some design. The former presumably occur continuously throughout life, are multiple, and are critical to the maturation and functional development of the adult masticatory apparatus. The latter, we may assume, modify original behavior patterns for a variable period depending upon the severity of the effect.

Experiments designed to measure alterations in masticatory behavior per se are few. Longitudinal studies, especially of mastication in the growing child are virtually nonexistent. This is particularly unfortunate, because it is here that profound learning changes occur through the deciduous, mixed and permanent dentitions. Eruption of teeth is a significant factor causing marked adaptational changes in the patterns of muscle contraction. Coupled to this is the developing child's increasing capacity for coordinated motor movements. Clearly we need to observe early attempts at mastication in the true sense, and the changes which occur as various teeth emerge, establish functional contact and when they are replaced.

In behavioral experiments involving the adult masticatory apparatus, emphasis has mostly been placed upon perturbations of the system, where specific interventions have been made in order to observe the behavioral response to them. The most popular experiments here have been to alter the occlusion for short periods by creating interferences which impede its normal function. Collectively the results have been equivocal, because different techniques have been used to study different variables. The findings can be summarized as demonstrating that when interferences are present, muscle responses (especially the peaks) occur earlier prior to intercuspation, associated with alteration in jaw movements, although not in a wholly predictable way [172].

Ethical considerations preclude long-term experiments of masticatory behavior. Yet even the alternative (which is ethical) of measuring the behavior response following conventional treatment has rarely been adopted. Given the known statistical correlation of skeletal and dental morphology with masticatory muscle function [178–181], it is surprising to find no systematic study of functional mastication before and after orthodontic therapy, orthognathic surgery or full-mouth reconstruction. This wealth of clinical material is like a Pandora's box waiting to be opened. Of particular interest would be the nature of brain changes following such marked alterations of the occlusal and skeletal pattern of the oral–facial region. One may predict that orthodontic therapy would force constraints upon the masticatory system through changes in the brain or structures within it (see Sect. 3.3, Chap. 4).

Occlusion and Mastication 7

7.1 Introduction

The pattern of occlusion of teeth is highly variable in the human, and in orthodontics there is an increasing tendency to attribute an abnormal pattern of masticatory movements as well as hyperactivity of the jaw-closing muscles to malocclusion of the teeth. For example, it has been suggested that in normal occlusion of the teeth chewing movements are simple and well coordinated, and the periodontal mechanoreceptors provide information reflexly to the trigeminal motor neurons to determine the muscle force and jaw movements required in each masticatory cycle [78, 93, 171]. On the contrary, subjects with malocclusion of the teeth may have an irregular, complicated pattern of chewing movements due to disruption of the function of the "central pattern generator" (CPG) which regulates the rhythm of the cyclic masticatory movements [78, 93]. There is now convincing evidence that specific patterns of malocclusion of the teeth can alter the symmetrical bilateral function of the masticatory muscles. For instance, while in normal occlusion the amplitudes of the responses recorded from the left and right masticatory muscles are very similar, in subjects with unilateral posterior crossbite malocclusion of the teeth there is asymmetry between the left and right masticatory muscles, suggesting abnormality of the bilateral chewing function [125], which may be due to abnormal coupling of hyperneurons of the left and right cerebral hemispheres (see Sect. 4.2).

The amplitude of an action potential is determined by the number of motor neurons which are activated by the stimulus. If the stimulus is constant, then the increase in the action potential that is recorded in the muscles of the unilateral posterior crossbite side, suggests an increase in the number of facilitatory impulses to the trigeminal motor neuron nuclei of the crossbite side, thereby increasing the number of motor neurons required in the response [125]. This increased excitability of muscles of the crossbite side may come from many sources. For example, the muscle spindles may signal to the brain that the jaw-closing muscles on one side of the head are a slightly longer, since the mandible is displaced from the midline of the face towards the crossbite side [125]. In the masticatory system the proprioception of the mandible (the sense of position in space of the mandible relative to the maxilla) arises primarily from the muscle spindles in the jaw-closing muscles, as well as from other mechanoreceptors in and around the mouth. Muscle spindle receptors are exquisitely sensitive to

DOI: 10.1007/978-3-642-00396-7_7,

stretch, and hence are able to signal the length of the jaw-closing muscles and, by inference, the vertical position of the mandible relative to the maxilla [78].

The phenomenon of uncoordinated masticatory movements in malocclusions of teeth has been attributed to the presence of occlusal interference or premature contacts of the teeth, which modify the activity of muscles through the receptors of the muscles, skin, and oral mucosa, as well as of periodontal receptors (see Chap. 8). When interferences are present, muscle responses (especially the peaks) occur earlier prior to intercuspation of the teeth, presumably through an increase in the excitability at least of the periodontal reflexes [172]. When there is displacement of the mandible during occlusion of the teeth, the muscle activity on the displaced side is altered, as for instance in the unilateral posterior crossbite malocclusion [124, 125].

Occlusal morphology plays some part in determining the precise pattern of movement in chewing, and therefore will determine the demands upon the different muscles, and upon motor units within the muscles. With a stable occlusion in the intercuspal position, it is reasonable to consider that the average work-load of individual motor units may remain constant over a period [78, 98, 173].

The presence of occlusal irregularities and premature contacts, however, has often been considered to be a key factor in the dysfunction of the mandibular muscles of both sides as a whole, and of individual motor units. Such circumstances may reasonably be regarded as producing local and/or central hyperactivity of muscles associated with abnormal movements. Although there is no direct cause and effect relationship between dysfunction and a defective occlusal pattern, it is reasonable to consider the possibility that the occlusal change initiates the functional difficulty, causing redistribution of muscle activity, which results in damage to the muscles. On the other hand, there is evidence which suggest that painful jaw muscle dysfunction occurs in spasm, implying excessive continuous contraction, which is likely to be associated with emotional states of anxiety [173].

However, the neuromuscular dysfunction associated with occlusal interferences or premature dental contacts may be related to sensory deprivation of the cerebral cortex through the affected ascending reticular activating system (ARAS), which regulates the sensory input to the cortex. This breakdown of the central reference of arousal of the brain for attention and awareness (a preconscious state of the brain, which correlates with the emergence of the "conscious electromagnetic field" instructing the muscles how to contract), by which the organism–mouth guides its correction strategies in perceiving, cognizing, reacting, and manipulating the environment, may make it increasingly difficult for the cerebral cortex to discriminate sensory inputs and to emit motor responses to muscles (see Sects. 3.4, 4.8).

7.2 The Chewing Cycle

Chewing is a cyclical activity of the jaw-opening and jaw-closing muscles occurring at a rate determined by the CPG. In chewing the CPG programs four phases of the masticatory cycle. The first phase is the preparatory phase, which begins with the opening of the mouth, through activation of the jaw-opening muscles, in order to receive food. The second phase is the food contact phase, which occurs by switching off the activity of the jaw-opening muscles and activating the jaw-closing muscles to produce the initial closing movement. It is in this phase of the cycle that the periodontal reflexes may assist in grasping the food in the correct position between the teeth, ready to be bitten through. The third phase is the food crushing phase, in which the output from the CPG to the jaw-closing muscles forces the teeth through the food bolus, producing the chewing strokes. In the last phase of the chewing cycle, the tooth contact phase, activation of the jaw-closing muscles continues as the opposing teeth come into contact, and while they slide into the intercuspal position for the final movement grinding the food into a paste. A reflex emanating from the periodontal mechanoreceptors initiates and controls this final grinding movement in the masticatory cycle [78].

7.3 Directional Sensitivity of the Teeth Guides Normal Occlusion

It has been suggested that the periodontal receptors are directionally sensitive and modulate the activity of the jaw-closing muscles to ensure that the final grinding phase of the chewing cycle occurs in the correct direction, i.e., from the working side towards the intercuspal position. The periodontal mechanoreceptors activated by contact of the teeth, signal to the brain that the correct pattern of contact of the opposing teeth has occurred before the grinding movement is reflexly initiated. In other words, if the cusps of the upper teeth do not meet the occlusal surfaces of the lower teeth in the correct relationship, the grinding movement is not initiated. This suggests that the central nervous system issues motor commands to the jaw-closing muscles, through the CPG, in proportion to the total periodontal afferent discharge evoked by the number of teeth in contact [78, 93]. At the end of the grinding movement the mandible is stabilized into the intercuspal position being under both mechanical tooth-guided control from the anterior teeth, and neuromuscular control mediated by periodontal input. At the end of the grinding movement the mandible slides into the intercuspal position. The CPG then switches off the activity of the jaw-closing muscles and reactivates the jaw-opening muscles to move into the next chewing cycle [78].

In this view, in a normal occlusion the "muscle position," meaning the optimum muscular function which guides the mandible to close in the intercuspal position without sliding in an eccentric position because of premature contacts, coincides with the "tooth position," which is the position of maximum intercuspation of the teeth, a reference position for normal occlusion of the teeth from which the coordinated masticatory activity can be initiated by activation of the CPG [173].

7.4 Functional Malocclusion

On the contrary, in a functional malocclusion of the teeth, during closure the first contact is unstable, and the mandible slides into an eccentric occlusal position. This indicates the presence of abnormal tooth contact (dental interference), meaning the absence of a stable occlusal reference position, which modifies muscle function. Accordingly, the "muscle position" and the "tooth position" do not coincide, resulting in no activation of the CPG. In such circumstances the masticatory activity is not coordinated, being under voluntary control, bypassing the CPG.

The following example illustrates more specifically the changes that occur in a functional malocclusion in which the lower first molars have overerupted. In this instance, the anterior displacement of the mandible during closure of the teeth may be induced by the overerupted molars acting as dental interference present in the central relation [174]. Accordingly, since in normal occlusion the preferred directional sensitivity of the periodontal mechanoreceptors of the lower first molar teeth is in a distal-lingual loading direction [97, 98], and given that in the above described functional malocclusion the sliding of the mandible is in an anterior direction following closure of the teeth, the lower first molars are likely to experience mesially directed forces, that is a direction opposite to that in which the periodontal mechanoreceptors of these teeth are most sensitive to discharge. It follows that the periodontal mechanoreceptors of the anteriorly displaced mandible will not encode information during chewing, resulting in altered masticatory activity.

In fact, it has been reported that anterior displacement of the mandible due to dental interference in the region of the lower first molar teeth reduces the chewing efficiency in the posterior part of the dentition, resulting in further overeruption of the molar teeth, which in turn causes more anterior displacement of the mandible followed by further reduction in the chewing efficiency. The vicious cycle continues in this manner leading to the development of a malocclusion of the teeth characterized by a prognathic mandible coupled with masticatory insufficiency [174].

7.5 Functional Malocclusion Affects the Central Pattern Generator

In a functional malocclusion with a slide of the mandible closing in an eccentric position in reference to the mid-line of the upper dentition, it has not been shown definitely which receptors are responsible for the deviation of the mandible. It has been suggested that proprioception in the limbs arises largely from muscle spindles [93]. It is then likely as already suggested that in the eccentric occlusion, the muscle spindles are signaling to the brain that the jaw-closing muscles on one side of the head are a slightly longer than on the other side. This implies that the functions of the left and right masticatory muscles are not symmetrical. The asymmetric function of muscles associated with eccentric occlusion of the teeth may suggest that during the last phase of the chewing cycle, the upper cusps on the left and right side do not contact symmetrically the occlusal surfaces of the lower teeth and hence the grinding movement, at the end of which the mandible is stabilized in the intercuspal position, is not initiated. The elimination of the last phase of the masticatory cycle indicates disruption of the cyclical chewing process. This may imply that the dental interference has affected the chewing rhythm of the CPG, possibly through sensory deprivation of the cerebral cortex.

7.6 Factors Affecting the Maturation of the Central Pattern Generator

The view that the occlusal interferences or premature contacts affecting the functional program of the CPG through sensory deprivation of the cortex (see Sect. 3.4) may be more relevant in early age during eruption of the teeth and their gradual replacement by the permanent teeth, since in young children the protective nervous system predominates over the discriminative system. In other words, in childhood the tactile stimuli are interpreted by the brain as nociceptive stimuli (see Sect. 8.30). On the other hand, several authors have suggested that the eruption of the teeth accelerates the maturation of the chewing function [1, 67, 78]. With the eruption of the teeth the masticatory movements become less reflexive and more under voluntary control, reflecting maturation of the cortical mechanisms, which accordingly program the CPG for cyclical coordinated movements. During these movements the force of muscular contraction is fine-tuned by the periodontal mechanoreceptors, as well as by the muscle spindles of the jaw-closing muscles [78, 93].

If, however, during development of the dentition dental interferences appear and the mandible is guided and/or in avoiding the interference occludes in an eccentric position, for instance in unilateral posterior crossbite, then, as described above, the muscles on the two sides of the head are unbalanced, and the eccentric occlusion of the teeth disrupts the function of the CPG. These changes in the function of the CPG may be the result of impairment or delayed learning of conscious masticatory function in the cortex caused by the sensory deprivation through the dental interference.

7.7 Form and Function

The chewing behavior varies very much in the human. It varies in terms of the shape of the chewing cycle, its sequence, the action of individual muscles, their temporal relationships to each other and to jaw displacement, and the size and nature of the forces produced between the teeth. The first major effect on variation in the chewing cycle is the anatomical substrate. People differ greatly in their craniofacial morphology. Thus, from both a biological and a clinical point of view, orthodontists and other dentists have been interested in the interactions between the muscle activity relative to changes in occlusal and jaw relationships. Some correlations observed between the morphological characteristics of the facial skeleton and the function of the masticatory system are reviewed.

Interest from a functional aspect regarding the form of the facial skeleton is directed mainly to the shape and size of the upper and lower jaws and the vertical and sagittal jaw relationships. Lindegard [178] and Björk [179] paid great attention to individual variations in the shape of the mandible in radiographic cephalometric studies of craniofacial morphology. They found that a mandibular shape characterized by a prominent upward bend of the mandibular base and a small gonial angle is seen unusually in individuals with the following features: a small inclination of the mandible in relation to the anterior cranial base and the nasal plane (the lower border of the mandible is nearly parallel to the anterior cranial base and nasal plane), a lower occlusal plane more or less parallel to the lower border line of the mandible, a small anterior face height in relation to the posterior face height and a tendency to mandibular prognathism, in other words a square facial type. In contrast, a mandibular shape characterized by a downward bend of the mandibular base and a large gonial angle is usually associated with the following features: a large inclination of the mandible in relation to the anterior cranial base and nasal plane, a marked inclination of the lower occlusal plane to the mandibular base line, a large anterior face height in relation to the posterior face height, and mandibular retrognathism.

In a study of the pattern of craniofacial associations by Solow [180], it was found that the height of the alveolar ridges of the anterior teeth is correlated positively with the inclination of the mandible relative to the nasal plane and the anterior cranial base. This association was interpreted as reflecting the dentoalveolar compensatory adaptation to the jaw relationship during growth. The type of occlusion, on the other hand, has been found to display great variability among facial types. Sassouni [181] reported that a skeletal deep bite is usually seen in subjects with a small gonial angle and a square facial type, while a skeletal open bite is usually seen in subjects with a large gonial angle and a retrognathic facial type.

The relationship between craniofacial morphology and masticatory muscle activity has received much attention. Studies in humans are few. Moller [182] in an extensive electromyographic investigation in young adults, related the activity of the masticatory muscles to facial morphology. He found that strong activity of the masseter and of the anterior temporal muscles in maximal clench is associated with a small inclination of the mandible to the anterior cranial base, a marked mandibular base upward bend, a small gonial angle and mandibular prognathism, in other words, with the square facial type discussed above. Conversely, in subjects with a retrognathic facial type, the temporal and masseter activity was found to be less marked. Similar correlations have been observed by other investigators. Further, an electromyographic study by Ahlgren [183] showed that a significant negative correlation exists between masseter activity during chewing and the size of the gonial angle.

In addition, Moller [125] found that in subjects with unilateral posterior crossbite malocclusion of the teeth, the mandible is displaced towards the side of the crossbite. Reference points and lines used to assess facial morphology on posteroanterior cephalometric radiographs are (a) the mid-point of the lower dental arch in relation to that of the upper and the facial mid-line ($p<0.001$), (b) the mandibular symphysis ($p<0.01$), and (c) the antegonion points ($p<0.05$) in relation to the facial mid-line. Displacement of the mid-point of the lower dental arch was present also on radiographs taken with the mandible at rest ($p<0.05$).

The displacement of the mandible is associated with predominance of the posterior temporal muscle on the side of the crossbite both with the teeth separated (rest position) and during maximal clench of the teeth in the intercuspal position. Ingervall and Thilander [184] have confirmed these observations, and in addition they found asymmetry of motor output between the left and right temporal muscles during mastication. Hence, a morphological deviation of the face particularly related to the occlusion of the teeth influences the distribution of activity in the posterior temporal muscles and thus the position of the mandible at rest and during mastication. Subjects with strong predominance of maximal activity in the posterior temporal muscle on the side of the crossbite have displacement of their antegonion points towards the side of the crossbite. This means that the morphological changes appear to be skeletal (osseous) in nature. The occlusal and skeletal deviation were attributed to cuspal interference and to structural asymmetry of muscle and bone, respectively, through functional adaptation. Fixation of muscle at shorter or longer lengths normally results in structural adaptation by a decrease or increase, respectively, in the number of sarcomeres, suggesting plasticity of muscles following neuroplasticity of cortical motor patterns. The dissociation of the left and right symmetric masticatory muscle activity in the unilateral posterior crossbite malocclusion of the teeth may suggest abnormal coupling of the hyperneurons of the left and right cerebral hemispheres (see Sect. 4.2). The extensive individual variation of muscle activity during chewing points strongly to variation of sensory information input coupled with the individual “central disposition” (see Chap. 4).

The statistical evidence that specific patterns of muscle activity during mastication and swallowing are correlated with the oral–facial morphology and occlusion of the teeth may be understood through the mechanosensation concept advanced recently by Moss [91]. Briefly this concept suggests that bone is electrically actively “hard-wired” with the action potential of nerves and that bone remodeling dynamics is tuned to the frequencies of the muscles (see Sect. 4.8, Chap. 2).

Finally, given the known influence of the skeletal and dental morphology upon chewing, it is surprising to find no systematic study of functional malocclusion before and after orthodontic therapy. The morphological changes that would result from orthodontic therapy would be expected to force constrains upon chewing behavior through brain changes (see Sect. 3.3, Chap. 4).

Applied Neurophysiological Concepts in Orthodontics

8

8.1 Introduction

Diagnosis of malocclusion of the teeth associated with evident neuromuscular dysfunction is a specialized professional skill requiring at least a basic knowledge of the mechanisms of the brain that underlie sensation, perception, learning, memory, and motor control of muscles. A thorough analysis of these mental events is presented in Chap. 4. It is the specific intent of this chapter to present certain neurophysiological concepts that could be used as adjunctive therapeutic modalities within the framework of a comprehensive approach geared to the management of oral neuromuscular dysfunction. Comprehensive management includes analysis of the oral–facial sensorimotor system and evaluation of physiological effects that the proposed biomechanical treatment could have on the oral neuromuscular behavior. The proper use of one or more adjunctive therapy modalities can significantly enhance achievement of the orthodontic treatment goals.

8.2 Oral Perceptual–Motor Dysfunction in Children

The oral–facial region may be regarded as a major sensory–perceptual–motor system of the body surrounded by a complex musculature [67, 154]. Stimulation of the sensory receptors of the skin and oral mucosa by experience gives rise to a patterned neural input and processing of information in the cerebral cortex, resulting in sensation, perception, learning, and memory of experience, as well as motor control of muscles and of salivary glands. Thus, the senses interacting with the environment give rise to a number of functions through transformations in the brain, depending on the stimulus [15, 38, 58, 85].

To interact with the environment, we must know about the environmental stimulation of sensory receptors and about attributing meaning to the response and the use of the stimuli by the brain. This point is so basic that it is easy to overlook it. Just because there is an environment acting on the child and that his/her receptors seem to be responding to stimulation, does not mean that the cerebral cortex is really interpreting the world in a meaningful

M. Z. Pimenidis, *The Neurobiology of Orthodontics*,
DOI: 10.1007/978-3-642-00396-7_8, © Springer Verlag Berlin Heidelberg 2009

manner, or with the same meaning that we might expect. One of the first things we need to learn to do is to avoid taking the functions of the cerebral cortex for granted [69, 71, 81, 82]. Perceptual or motor functions are not an either/or matter, but one of degree, just as any ability varies by degrees. Because the motor acts are so closely associated with perception of our mouth structure and because we so often express our interpretations of our tangible environment in motor acts, the perception and motor processes must be considered together. This reasoning accounts for the use of the term perceptual–motor [63, 69, 85].

Let us further analyze the behavior dimension we are discussing. Let us begin with the motor component. We know how the mouth is constructed (we are talking about the mechanical aspects) and how it moves, by seeing it and feeling it. The major types of sensory receptors through which this information comes are visual, tactile, proprioceptive, and vestibular. Although the vestibular sense is proprioceptive, because it is a specialized sense it is considered separately from other proprioception. When we refer to proprioception, we are referring to all those sensations, which arise from muscles, tendons, joints, fascia, and related tissues. It is noted that taste is not a pure sensory modality, but rather a mixed sensation composed of gustatory, tactile, pressure, cold, heat, and olfactory impressions. Taste is not a pure modality because the tongue perceives not only sweet and saline, but also the weight, fluidity, roughness or smoothness, temperature, viscosity, and volatility of food [42].

Our task then in orthodontics is to understand the nature of the central nervous system's mechanisms, which integrate the oral–facial sensory input and create the appropriate output. The more careful the evaluation of these essential processes and the better the insight we have of the perceptual–motor problem, the more effective will be the treatment. An orthodontist who seeks treatment procedures only, without a firmly based understanding of the sensorimotor dysfunction requiring treatment, is advised not to treat at all.

For example, in unilateral posterior cross-bite malocclusion of teeth, there is dissociation of the normal bilateral activity of the masticatory muscles [124, 125]. The patients, however, do not seem to perceive a difference between the activities of the left and right masticatory muscles. It is interesting that the orthodontic literature gives very little account of how this central sensorimotor–perceptual dysfunction affects the bilateral firing (and contraction) of the masticatory muscles.

Recent neurophysiological studies, however, suggest that thalamocortical neurons generating synchronous alpha and theta cortical activity in the electroencephalogram (EEG), are linked by gap junctions (electrical synapses) in the thalamus forming a giant neuron or hyperneuron [123], so that thalamocortical projections (or trans-corpus callosum pathways) could couple both hemispheres in hyperneurons to account for bilateral synchronous activity [38]. Gap junction-connected neurons are the cellular level of the neural correlate of perception or consciousness. A key feature of gap junction-connected neurons is continuity of dendritic membranes of neurons that depolarize coherently. Another key feature is continuity of cytoplasmic interiors [38].

Thus, abnormal coupling of cerebral hemispheres in hyperneurons seems to underlie the dissociation of the activity of the left and right masticatory muscles in unilateral posterior cross-bite malocclusion of teeth. This might suggest that the normal bilateral masticatory function has never been consciously learned and hence no memory record for bilateral function of muscles has ever been inscribed in the left and right motor cortices. This relationship between perception and cognition is very well established [15, 58]. In fact, motor learning and perception form a continuum [85].

Judging by those aspects we have considered so far, we may assume that we are concerned with the child's defective manipulation of his sensory environment. In this view, the most obvious and significant changes in the electrical and chemical activity of the brain occur not necessarily with specific oral stimuli, but are dependent upon the meaning of the stimuli to the child's cerebral cortex [69, 82]. If the stimulus is to warn the brain of danger and to help mobilize it for defense by producing a high level of arousal for alerting the brain (fight or flight response) then the very cortical processes that are essential for perception and learning are inhibited. Thus, over-arousal of the brain impedes discrimination of senses, resulting in perceptual deficit and inability to learn the experiences provided by the senses [63, 64]. This means that if during a critical period of development a child complains of oral–facial pain, tooth ache, or another oral disorder causing over-arousal of the brain, then learning of new experiences may be inhibited.

Similarly, if the central nervous system is interpreting the orthodontic treatment stimuli (tactile, pressure, proprioceptive, etc) as danger signals, then the ability of the cortex to discriminate and process these stimuli will be suppressed, resulting in no control of the expected orthodontic changes. The implications of these cortical reactions in orthodontic practice are obvious. We must try to provide the kind of sensory and motor experiences to patients that will normalize the central nervous system's ability to process the sensory input that the treatment provides. In other words, if we can inhibit the over-protective response of the brain, which calls for fight or flight reaction, then we will likely facilitate discrimination of sensory–perceptual processes, learning and memory abilities, and motor control of muscles. Thus, the important point is to inhibit the irrelevant stimuli. Our objective is to accomplish this on a neurophysiological level, the way the normal brain does it, rather than by simply preventing the stimulation of receptors. The reticular formation and the ascending reticular activating system (ARAS) play a critical role in inhibiting distracting information, as well as in facilitating the relevant stimuli reaching the cerebral cortex, enhancing then learning and perception ability [64].

Survival value is one of the strongest principles of sensorimotor functions. The protective system predominates in early life and often whenever development deviates from the normal. Predominance of the protective system inhibits manifestation of the discriminative system. Activation of the discriminative system inhibits the protective system. Maturation inhibits the protective system and enhances the discriminative system, probably through maturation of inhibitory nerve fibers and inhibitory mechanisms of the brain [79].

8.3 The Scheme of the Mouth

The question that arises is what type of development must occur before the mouth of a child is able to react effectively to environmental stimuli. An obvious answer is that the child must learn just how his/her mouth is constructed and just how it moves. In other words, the child must build an adequate oral scheme. The word "learn" does not refer to cognitive learning. Little is to be gained by teaching a child with oral dysfunction that he/she has, for instance, twenty teeth, in the same way that you would teach in the same child the concept "twenty." The scheme of the mouth is not acquired in that manner. It is acquired through the sensory receptors and, it is hypothesized, through reinforcement from the results of purposeful movements. If the movement accomplishes some desired result the brain "notes" that certain directions are effective in bringing about certain movements, and that the same directions may be used again in a comparable situation. For example, the balanced action of the paired genioglossus muscles is required to protrude the tongue straight out of the mouth (the base of the tongue is pulled forward by the simultaneous contraction of the genioglossus muscles). If one genioglossus is inactive the action of the intact muscle is unopposed, resulting in deviation of the tongue towards the inactive muscle, suggesting damage to the hypoglossal nerve (lower motor neuron lesion).

Research in apraxia (perceptual–motor disorder or syndrome affecting especially the skilled motor functions) suggests that the role of receptors deserves particular note. While one might assume that the kinesthetic (proprioceptual) receptors such as muscle spindles might be critical for providing information about their position and movement, scientific data force us to give even more attention to the tactile receptors in the development of the ability to motor plan. It is the tactile receptors that provide us with much of the information about the nature of the external world that we are manipulating with our fingers and tongue [93]. We all know how hard it is to handle coins with gloves or to put lipstick on after a Novocain shot from the dentist.

We must be careful to distinguish here between mere awareness of sensory stimuli and the ability to interpret those stimuli, especially as to their spatial elements. If we touch a child's cheek when he is not looking and he indicates that he felt the stimulus, his response in no way indicates that he knew where he was touched. Discriminative interpretation of the spatial and temporal elements of tactile stimuli are essential to the development of the scheme of the mouth. Note that there is a difference between the concept of the scheme of the mouth and the idea that one needs tactile information at the same time as execution of the motor task. While the latter is an important process, motor planning, as an antecedent of execution, is dependent upon meaningful organization of previous tactile experiences. The scheme of the mouth is a neurological organization of previous tactile, proprioceptive and probably vestibular stimuli in association with planned movement.

The major questions facing us now are what kind of developmental processes must occur to assure the perception of the spatial elements of tactile stimuli, and what central nervous system mechanisms might be responsible for these processes? To answer these questions it is helpful to conceptualize the tactile functions of the nervous system. It has been suggested that there are two tactile systems: one designed primarily for warning the organism's mouth against impending danger and one designed to convey information about the environment for better interpretation of the danger. The two systems can be called "protective" and "discriminative", respectively. Their existence is finding increasing support from current research. It is the adequate function of the discriminative tactile system which is required for praxis or motor planning [79, 164].

Survival value is one of the strongest principles of sensorimotor function. The protective system predominates in early life and often whenever development deviates from normal. Predominance of the protective system inhibits manifestation of the discriminative system.

Activation of the discriminative system inhibits the protective system, most likely through maturation of the inhibitory nerve fibers and inhibitory mechanisms [37, 64].

In this context, we might need to advance new methods to improve oral perception in cases of oral–facial neuromuscular dysfunctions associated with some form of oral agnosia, namely through enhanced stimulation of the tactile receptors in such a way that it will inhibit the protective system and facilitate the discriminative system. Stimuli should precede and be used in conjunction with motor tasks.

For example, there is good neurophysiological evidence that tactile stimuli from the face, lips, and oral mucosa can be activated during voluntary movements of the jaw-closing muscles, through deformation of these tissues, suggesting that the proprioceptive signals of tactile receptors may be specific to the control of the opening and closing rhythm of the mouth during chewing [23, 78]. Some contraindications for using tactile receptors to enhance oral perception and movement are the presence of hypertonicity (muscle spasm), pain, and a strong negative reaction on the part of the patient. The younger the patient, the more careful we need to be.

8.4 Oral Sensory and Motor Somatotopic Coincidence

Another neurophysiological detail to be considered is that in the sensorimotor cortex the oral tactile and motor patterns coincide somatotopically. This means that the tactile impulses that arise when one holds an object between the teeth go to the part of the cortex that when stimulated causes motion of the mouth. A reasonable explanation of somatotopic coincidence would be that the distinction between the motor (precentral gyrus) and the sensory (postcentral gyrus) areas in the cerebral cortex is not clearly delineated. Both the precentral and postcentral gyrus are actively involved in sensory and motor processes [36]. This anatomical arrangement in the cortex may suggest that the oral tactile stimuli facilitate the discharge of the cortical motor neurons that are involved in the motion of the mouth, and vice versa that movements of the mouth stimulate the tactile receptors of the mouth. Thus the therapeutic scheme involving tactile stimulation in conjunction with motor tasks may be justified on the basis of oral sensory and motor somatotopic functional coincidence. The cerebral cortex, of course, is considered the major mediator of oral skilled motor functions that are so vulnerable to cortical dysfunction of "agnosia of teeth" that is discussed next.

8.5 Agnosia of Teeth

The density of the tactile innervation of the periodontal ligament and of the oral–facial region in general and its possible implications for tongue posture and the position of the teeth (mechanisms that are described later in this chapter) is reflected in the enormous size of the trigeminal and facial nerves and their large representation in the primary somesthetic cortex. Roughly half of the somesthetic cortex processes information from the oral, facial, periodontal, and pharyngeal regions. The large oral–facial and pharyngeal representation in the cerebral cortex explains our heightened awareness of the face, mouth, tongue and teeth, and why agnosia of these structures, such as agnosia of teeth (the brain is not aware of the presence of the teeth), is one aspect of inadequate oral perceptual–motor scheme disorder or oral apraxia, as discussed below.

A 21-year-old male patient with severe class III open bite malocclusion of the teeth who was not aware without looking of the presence of the incisor teeth in his mouth has recently been reported. The agnosia of teeth was associated with oral motor dysfunction including tongue thrust and impairment of coordinated chewing movements. However, after closure of the open bite and the establishment of normal occlusal and skeletal relationships through orthodontic therapy the patient became conscious of the full complement of his teeth [87]. Accordingly, it has been suggested that normal occlusion of the teeth is essential in forming the basis of normal perceptual awareness of the scheme of the mouth (see Sect. 3.3). Clinical studies support this conclusion and suggest that normal occlusion of the teeth is necessary to maintain the complete normal sensibility of the teeth. In an individual with malopposed teeth sensibility is more or less disturbed, and the individual will have a poorer ability to discriminate the physical characteristics of objects held between the teeth [88].

The role of receptors involved in oral agnosia deserves particular note. While one might assume that the kinesthetic (proprioceptive) receptors, such as muscle spindles, might be critical for providing information about position and movement, scientific data on limb apraxia suggest more attention to the tactile receptors in the development of motor planning ability. It is the tactile receptors that provide much of the information about the nature of the external world that we manipulate with our fingers and tongue [93]. The tactile receptors of the finger tips have neurophysiological properties similar to those of the tactile receptors of the tip of the tongue [93, 157]. It is mainly the tongue tip that is used to manipulate and explore objects and structures in the mouth, including the teeth. It is then most likely that it is the tactile receptors of the tip of the tongue, as well as the periodontal mechanoreceptors, that are affected in agnosia of teeth. It is noted that all the basic five sensory systems of the mouth need to be working simultaneously and cooperatively for higher skill acquisition and performance. In fact, if any one system does not work properly either by itself or in conjunction with others, a sensory dysfunction of some kind may result [79].

Thus the discriminative interpretation of the spatial and temporal elements of the tactile stimuli at the tip of tongue and of the periodontal ligament are essential for the development of the scheme of the mouth, including awareness of the presence of the teeth. Note that the scheme of the mouth is a neurological organization of previous tactile, proprioceptive, and probably vestibular stimuli in association with planned movement [154].

8.6 Oral Apraxia

As orthodontic specialists dealing with neurophysiological problems of oral motor dysfunctions, we study in greater depth the central nervous system's function in order to have a better understanding of the perceptual–motor problems of the masticatory system. Oral apraxia is one type of oral perceptual–motor disorder or syndrome. It is a disorder of skilled motor function—a disorder of the ability to plan coordinated masticatory movements of the mouth, as opposed to a disorder of executing the motor plan, which involves a lesion of the neurons of the precentral motor cortex or their axon channels (upper motor lesion).

Some degree of developmental oral apraxia should be suspected in, for example: an orally clumsy child with uncoordinated chewing movements; a child with dissociation of the left and right masticatory muscle function (as occurs in unilateral posterior cross-bite malocclusion of the teeth), with difficulty in jaw and tongue postural position or with difficulty in recognizing the shape of objects put in the mouth without the use of vision; a child who is not aware of the presence of teeth in the mouth without the use of vision; a child with disordered taste or salivation; a child who is a picky eater (dislikes certain textures, tastes or temperatures); a child who does not know when his mouth is "full"; a child who vomits or gags easily; or a child who is clumsy with fine motor activities (eating, drinking, etc) [79]. The condition is especially evident when the child is faced with a new oral task which is unfamiliar and which requires considerable focus of attention for learning, as for instance touching the nose with the tip of the tongue.

The clumsy child who has difficulty learning to manipulate objects in the mouth could be helped with tactile and pressure stimuli in conjunction with motor tasks followed by assumption of postural patterns. These procedures may activate the discriminative system while at the same time inhibiting the protective system. Remember that in the ontogenetic process, the child first learns to plan gross motor actions, gradually refining his/her skill until, for instance, he/she writes with ease and bites and chews with finesse. Most of us forget that we all once motor planned walking and sitting in a chair. We no longer have to plan them.

For a task to be therapeutic the child must have some success in the task which should demand that he/she motor plan. Any gross motion frequently repeated, such as walking, does not make a demand for planning. Activities on which we must focus our attention do require planning. This means that the stimuli must cause arousal of the brain for attention and awareness which correlate with conscious learning, memory, and motor planning (see Chap. 4). Thus walking backwards or putting the tongue in specified spots in the mouth may cause arousal of the brain for learning new experiences. Motor planning should be taught, taking the child passively through the motions, for it is the sensory stimuli we learn when we learn to motor plan.

In addition, crossmodal influences, which appear instantly and disappear rapidly during early development in the primary and secondary cortices serving multisensory tasks, may be involved in learning ability [15]. For example, a recent functional magnetic resonance imaging (fMRI) study has identified a region of the brain which is activated specifically during a combined visual–auditory task [134]. In another study neurons in the secondary auditory cortex tuned to precise frequencies of sound also responded to a tactile input [135]. Similarly, it has been found that visual stimuli enhance tactile activity through modulation of responses in primary and secondary somatosensory cortices serving multisensory components [15].

In this view, oral stereognosis (the recognition of form of objects put in the mouth, without the benefit of vision) is a multisensory experience; it involves at least the integration of an oral kinesthetic (proprioceptive) sensory modality providing information about the position in space and the tactile and pressure receptors providing information on the form of objects, and can be enhanced with the conscious motions of the mouth in early age. One of the directions of evolutionary development has been intersensory experience, that is multisensory integration coming from the same sensory system, for instance the mouth, and crossmodal experiences involving the integration of two or more sensory systems. Crossmodal experiences involving visual–auditory–tactile information are involved in the development of speech (see Chap. 5).

8.7 Enhancing Oral Perception: Pacini Receptors

The orthodontic practice is making demands for inclusion in our treatment of oral apraxia, the development of the discriminative and learning ability of the mouth, through the motor system, which activates in addition to other receptors, the Pacini corpuscles, one of the major proprioceptor mechanoreceptor, known to send information to the cerebral cortex for conscious perception of the moving tongue and jaw. These receptors are found in the masticatory muscles and temporomandibular joints. The Pacini corpuscles are rare or lacking in most areas of the facial skin and oral mucosa, the ventral side of the tongue being a noted exception. These sensory endings have a higher threshold to stimulation than many other proprioceptors. This suggests that heavy muscular exertion is more beneficial in activating the Pacini corpuscles, which convey the modalities of pressure (deep sensibility), stereognosis and of course proprioception. The Pacini corpuscles respond to muscle stretch and contraction, as well as to the movements of the temporomandibular joints [44, 93]. The use of heavy class II intermaxillary elastic forces in orthodontics to protrude the mandible also apparently leads to stimulation of the Pacini receptors of the patient's muscles and temporomandibular joints.

It has been suggested that the Pacini corpuscles of the temporomandibular joints play an important role in the conscious perception of the angular position of the mandible [162, 163]. The ability of the Pacini corpuscles to monitor mandibular angular position is impaired by severe malocclusion of the teeth with a reduced anterior vertical dimension of the lower face [169]. These studies might substantiate the orthodontic restoration of the vertical dimension of the lower face in patients with deep overbite [169]. In this view, mandibular angular proprioceptive perception and motor control could be enhanced through muscle contraction, especially postural patterns, against resistance, stimulating the high-threshold Pacini corpuscles in the masticatory muscles and temporomandibular joints.

Functional jaw orthopedic appliances which reposition the mandible to alter the muscle forces against the teeth and craniofacial skeleton may stimulate the Pacini receptors in addition to muscle spindles. Functional jaw orthopedic appliances are used where neuromuscular dysfunction has played a role in the etiology of the malocclusion and/or where enhancement or alteration of normal functional activities may provide optimal conditions for growth and development of the jaws and occlusion of the teeth.

8.8 Enhancing Oral Perception: Muscle Spindles and Tactile Receptors

Another approach which might eventually be more effective than facilitating the perception of moving oral structures lies in the function of the muscle spindles. However, there seems to be little neurophysiological support from these receptors in the oral–facial region. There is little evidence to indicate cortical receipt of information from the spindles of jaw-closing muscles. The jaw-opening muscles have no spindles [78]. Therefore, in order to understand the proprioception of the moving mandible relative to the maxilla we need to extrapolate from studies of the stretch reflex in spindles of the limb muscles.

The proprioception of the position of moving limbs relative to other structures of the body arises largely from muscle spindles, although inputs from tactile mechanoreceptors of the skin that covers the muscle can be even more effective, particularly in the fingers and in the oral–facial region. These tactile receptors provide discriminative interpretation of the spatial and temporal elements of tactile stimuli, which is essential for the development of the ability to plan the movement of the mouth. In fact, it is the tactile receptors of the skin and of the oral mucosa that provide much of the information about the nature of the external world that we experience with our hands, mouth, and body [78, 93].

There is good neurophysiological evidence that tactile stimuli from the face, lips, and oral mucosa may be activated during voluntary movements of the jaw-closing muscles through deformation of these tissues, suggesting that the proprioceptive signals of the tactile receptors may be specific to controlling the opening and closing rhythm of the mouth during chewing [23, 78].

The effect of tactile stimulation may have a direct effect on the transmission of kinesthetic (proprioceptive) impulses to the cerebral cortex. The effect may come through centrifugal influences on the kinesthetic receptors and their relay stations. Another mechanism through which tactile stimuli enhance kinesthetic and visual perception is through the ARAS [64].

8.9 Enhancing Oral Perception: Centrifugal Influences

This seems an appropriate place to introduce another neurophysiological process which undoubtedly has a profound influence on perception, but the real functional role of which is little understood at present. It is not only the external environment which acts on the sensitive receptors of the oral–facial region, but also the central nervous system itself. Through descending axon channels, the brain influences the flow of sensory input at each relay station and at each sensory receptor [64]. The principle is seen and understood most clearly in reference to the muscle spindle. One of the most effective methods of influencing motor function is by controlling the sensory input from the muscle spindle [78]. The sensory input in turn is influenced by the motor task confronting the patient, in accordance with the principle that most of the oral sensory input is generated principally by the motions of the mouth and vice versa, i.e., the sensory stimulation generates the motion of the mouth [43]. The centrifugal influences serve the purpose of making the receptors and pathways more capable of sending information or preventing the normal deluge of sensory impulses from reaching the cortex. Thus, selection of the incoming sensory information in the central nervous system seems to be the basic principle [82] (see Sect. 2.6).

8.10 Affecting Oral Perception and Learning

The ARAS and its anatomically and functionally close associate, the diffuse thalamic projection system, are concerned with arousal of the brain for attention and learning purposes. The ARAS allows selective attention, augmenting meaningful and inhibiting irrelevant information. Sometimes, however, the purpose is to produce a high level of arousal or to alert the brain to danger and help mobilize it for defense (fight or flight reaction). In this instance, the very cortical process that is essential for perception and learning is inhibited because the brain's priority is the danger [63, 64].

For example, if the central nervous system is interpreting a tactile stimulus (and perhaps other stimuli, such as an auditory stimulus) as a danger signal, then at least two major mechanisms will affect perception. The tactile perception essential to motor planning will be depressed, and the ability of the nervous system to discriminate, separate out, augment, or inhibit stimuli will be reduced. If we can inhibit this overprotective response of the brain, then we will likely facilitate the perceptual process. The only method for which we have any assurance of success is through tactile stimulation accompanied by pressure (deep sensibility).

Thus, the reticular formation plays a role in inhibiting distracting information input to the brain. It also plays the companion role of increasing the information input to enhance perception. Activation of the primary, specific sensory pathways alone, such as the spinothalamic or medial lemniscal pathway, is not a sufficient process to enhance perception. In addition to receipt of information over these specific pathways, perception requires generalized nonspecific activation of the cortex from the diffuse effects of the reticular formation. All five sensory modalities contribute to the activation of the diffuse system through crossmodal stimulation. For instance, visual stimuli enhance tactile acuity through modulation of responses in the primary and secondary somatosensory cortex [15, 136].

A general therapeutic procedure which occupational therapists use to enhance perception of any sensory modality is to precede a specific perceptual task with general stimulation (tactile, proprioceptive, vestibular, olfactory, etc) which, however, is not such as to cause over-arousal and the protective defense response of the brain, but is to be received by the central nervous system as a comfortable and pleasant stimulation. The general arousal effect can be expected to last for a number of minutes.

8.11 Affecting Oral Perception: Emotional Influences

Another vital point concerning the centrifugal influences in oral sensory–perception–motor functions lies in the fact that each sensory receptor is also served by an autonomic system nerve. The discharge of muscle spindle receptors and cutaneous receptors is increased by sympathetic nerve stimulation. The sympathetic nerves exert a facilitatory effect on reflexes arising from afferent stimulation, including the enhancement of muscle contraction [93]. On the other hand, the sympathetic system is responsive to emotions. Accordingly, since perception is related to cognitive (learning ability) and emotional development [63] whose mutual developmental depends upon adequate patterned sensory input [69], it is not unlikely that the degree and kind of emotional and cognitive involvement manifested by a child will, through centrifugal influences acting on sensory reception and transmission, have a direct and specific effect on the development of perceptual processes [69]. This reasoning leads to the assumption that the motivated child will receive more benefit from the orthodontic therapy than the nonmotivated one. Our goal, however, in emotional involvement is not merely to elicit cooperation or to provide a pleasant experience during the therapy, but to alter the sensory–perceptual–motor function of the oral–facial region through the appropriate response of the cortex.

Under certain conditions, which we will assume to be associated with brain dysfunction, the balance of the protective and discriminative system is tilted in the direction of protective system predominance. In this instance, centrifugal influences lower the threshold of the oral receptors and the sensory impulse transmission favors a nociceptive interpretation of tactile stimuli. Specifically, it is light touch stimuli that are most apt to elicit the negative, defensive response. What protective system predominance does to the affect is only conjecture. It is suggested that we watch for elements of a basic fighting response or one of fear and flight. Heightened annoyance, "unreasonable" anger or belligerence may appear. Some of the elements of the physiological basis of the emotional reaction may be seen in the responsiveness of the brainstem reticular activating system to adrenaline. It is to this level of the reticular formation that is delegated the responsibility for alerting the organism to potential danger. Fear produces adrenaline and adrenaline not only enhances readiness for fight or flight but also lowers the threshold of some tactile receptors.

Thus, here we see the possible neurophysiological basis of the effect of emotions on perpetual–motor function. There is undoubtedly an interaction effect on a neurophysiological level between perception and emotions. The nervous system geared to fight or flight is at the same time inhibiting the very cortical recruitment process essential to perception and learning. This level of perception and learning is not restricted to the highly cognitive levels. Depression of the perceptual process relating to somatic (mouth) factors also likely occurs.

This theoretical formulation provides us at least with a direction for a physiological basis for orthodontic treatment. Instead of simply shielding a child from distracting stimuli, let us aim toward providing the kind of sensory and emotional environment which will encourage a better balance between protective and discriminative systems. It is suggested that we try pressure touch as opposed to light touch cutaneous stimulation, applying the stimulus especially to those parts of the body richly supplied with tactile receptors. These areas are hands, face, and forearms [93], and are discussed next.

8.12 Proprioception of Muscles of Facial Expression

The muscles of facial expression are also involved in proprioceptive perception. These muscles, however, lack proprioceptive mechanoreceptors, such as muscle spindles or Golgi tendon organs. Instead, the proprioceptive function of these muscles is served by tactile mechanoreceptors, which are embedded in the skin, lips, and oral mucosa. The muscles of facial expression, innervated by the facial nerve, insert directly into the skin of the face and their contraction deforms the skin in very specific patterns, causing the tactile receptors to discharge vigorously in a way that can be interpreted meaningfully by the central nervous system. The precise nature of the neurophysiological response of the tactile mechanoreceptors indicates that in addition to touch and pressure response, they also serve as proprioceptors, signaling detailed information about the consequences of activation of the muscles of facial expression in the perioral and lip tissues. Thus, the action potential of the tactile mechanoreceptors in the oral–facial region, elicited through deformation of the skin during voluntary movement of the muscles of facial expression, is particularly important for the proprioceptive control of these muscles [93, 155, 156].

Surface electromyograms from the lips during mastication suggest that the lips are activated most vigorously during the opening phase of the chewing cycle (primary activity) in order to produce a complete or partial anterior oral seal preventing the food from leaving the oral cavity. This is a consistent finding of electromyographic studies in subjects of all ages. Fine wire electrodes have localized this activity to the oral sphincter, the orbicularis oris muscle. Whether the primary activity of the lips results in a complete anterior seal or merely prevents the lips from being separated as much as the jaws is of little importance. From a neurophysiological point of view the significant finding is that the activity causing the lips to approach increases with morphological traits that tend to impede lip closure. For example, one of the most frequent abnormal lip functions which tends to impede lip closure is tongue protrusion in swallowing. The mylohyoid muscles act strongly when the tongue protrudes by raising the floor of the mouth and stabilizing the hyoid bone [170].

The degree of primary activity in the lower lip during mastication is related to facial prognathism. In subjects with retrognathism of both jaws the lips are usually sufficient and the anterior seal demands slight activity; with prognathism the lower lip may be taut at rest and chewing requires strong activity. The relationship between lip function and facial morphology exemplifies an important feature: the lips adjust the oral cavity to the particular function being performed (mastication, swallowing, speech) and are primarily activated to produce their own movements. The adaptive function seems to depend more on recognition of the shape of the supporting hard tissues than on the tension produced. The tactile mechanoreceptors of the perioral skin are in keeping with this assumption [170].

The lips are also active during the closing movement of the jaw (secondary activity). Recordings made with wire electrodes from individual muscles moving the mandible (for example, the right and left digastric, the right and left posterior temporal, the right and left lateral pterygoid) show that the secondary activity in the upper lip originates from its upper levator muscle for the purpose of withdrawal. The activity during jaw closing is especially pronounced in subjects with deep overbite and retroclination of the upper incisors, reflecting a more acute demand for active withdrawal of the upper lip during closure [125].

In malocclusions of the teeth associated with upset of the oral seal (incompetent lip function), the deformation of the perioral skin during speech, chewing, swallowing etc, may not be specific, resulting in inadequate activation of tactile mechanoreceptors and a deficit in their proprioceptive input to the brain, followed by impairment of the perceptual and motor function of the perioral muscles, for instance the orbicularis oris muscle.

8.13 Neurophysiological Similarities of Tactile Oral–facial and Skin Receptors

Recently it has been suggested that the tactile mechanoreceptors that supply the human perioral skin, lips, and oral mucosa have neurophysiological properties similar to those of the tactile mechanoreceptors that supply the human hand and arm. Similarly, the neurophysiological properties of the tactile receptors of the tongue tip are similar to those of the receptors of the finger tips. Thus, the similarity in the sensory innervation between the oral–facial region and the fingers is well suited to the functional specialization of these areas. The tongue tip and finger tips serve to manipulate and explore objects in the mouth and by the fingers, providing the oral and manual stereognosis experience, respectively. The perioral tactile mechanoreceptors, those of the lips and mucosa, and those of the skin of the hand and arm serve to signal the movements of the muscles of facial expression, masticatory muscles, and the fingers, respectively, through deformation of the tissues by contraction of the muscles that stimulate the embedded mechanoreceptors [93, 155, 156].

Specifically, four functionally distinct types of mechanoreceptors have been described for the glabrous skin of the hand. These four types can be distinguished by the following properties: (a) the rate at which their discharge activity adapts to sustained suprathreshold forces, and (b) the size and borders of their receptive fields. Two types adapt quickly (fast-adapting, or FA type I and FA type II) to sustained skin deformation, i.e., they respond only to the onset of a stimulus. The other two types (slowly adapting, or SA type I and SA type II) adapt slowly and signal the magnitude of sustained skin deformation [93, 159].

The type I (FA I and SA I) mechanoreceptors have small receptive fields, typically 2–8 mm in diameter on the finger tips, with well-defined borders. These properties and the high densities of these receptors on the finger tips (about 250 receptors per square centimeter) make type I mechanoreceptors particularly well suited for encoding highly detailed spatial information, e.g., for extracting information about the fine form and texture of an objects that is touched and manipulated with the finger tips. In contrast, the type II (FA II and SA II) mechanoreceptors are fewer in number and have receptive fields that are larger with less well-defined borders. Typically, their receptive fields cover a whole finger or much of the palm. The very sensitive FA II mechanoreceptors are easily excited by high frequency mechanical oscillations produced by a tuning fork held on the skin. The SA type II mechanoreceptors readily respond to lateral skin stretch produced by stimuli both near and at a distance from the area of the receptive field [93, 157, 159].

These observations indicate that the different types of mechanoreceptor extract information about different aspects of mechanical deformation of the skin, which is determined by the manner in which each receptor terminates in the tissues. It is thought that each of the four different types of mechanoreceptor terminate in morphologically specific nerve endings: FA I are Meissner corpuscles, SA I are Merkel cells, FA II are single Pacini corpuscles, and SA II are single Ruffini endings [93].

In general, the mechanoreceptors that supply the human facial skin, lips, and oral mucosa have properties similar to mechanoreceptors that supply the human hand. Specifically, the receptive fields innervated by the infraorbital nerve are characterized by rapidly adapting mechanoreceptors resembling the FA type I of the glabrous skin of the hand, as well as by slowly adapting mechanoreceptors with a high dynamic sensitivity (i.e., they respond vigorously to the onset of the mechanical stimulus) and an irregular discharge during maintained tissue deformation, resembling the SA type I receptors of the glabrous skin of the human hand [93, 155].

In the receptive field innervated by the mental nerve there is another group of mechanoreceptors characterized by a slow regular discharge rate and spontaneous activity, which are similar to SA type II mechanoreceptors of the skin of the hand [93, 155].

In the tongue tip the receptive field is innervated by the lingual nerve. Most mechanoreceptors terminating superficially in the tongue have extremely small and well-defined receptive fields and adapt quickly (FA) to maintained tissue deformation. The sensory innervation of the tongue tip resembles the innervation of the finger tips [93, 155].

The high density of tactile mechanoreceptors with small and well-defined receptive fields at the tip of the tongue and the lips is reflected in the spatial resolution acuity of these areas. For example, testing the two-point discrimination suggests that the area of highest tactile acuity of the body is the tip of the tongue with a threshold of 1 mm, followed by the lips and the tip of the fingers. The mechanical compliance, i.e., the ability of the oral–facial tissues to deform to touch, is high, and it is exemplified by the delicate perinasal skin and lip regions. The minimal mechanical stimulus that can be detected in the perinasal skin and lip regions is as low as a weight of 5 mg, while the threshold for the finger tip is a weight in the range 40–100 mg [93].

8.14 Periodontal Tactile Receptors

Another important receptor mechanism that contributes to motor control of the jaw-closing muscles is through the periodontal tactile mechanoreceptors. These receptors provide conscious information about the load applied to each tooth that is essential for the fine regulation of muscular force during chewing and occlusion of the teeth. The periodontal mechanoreceptor also signal the presence of objects between the teeth, the presence of high spots or premature tooth contacts during occlusion the of teeth, as well as the biomechanical forces applied to the teeth during orthodontic therapy. In addition, the periodontal mechanoreceptors are specialized to provide information about the deformation of the tissues during chewing, and serve also a proprioceptive role in the specification of the positions and movements of masticatory structures during function. This information is collected and processed in the central nervous system and is used to regulate subsequent muscle activity. In doing this the central nervous system is predicting from past experience of the chewing cycle to regulate the current chewing cycle. Thus the sensory experience gained in each preceding chewing cycle helps to determine the muscle force and jaw movements required in the next chewing cycle [93].

The sensory nerves that innervate the periodontal receptors have their cell bodies located in two sites, namely the trigeminal (Gasser) ganglion and the mesencephalic nucleus of the trigeminal nerve. Neural activity originating in the periodontal mechanoreceptors is relayed through fast conducting large myelinated axon channels in the trigeminal ganglion which projects to the main sensory nucleus and the spinal nucleus of the trigeminal nerve. The action potential from these sites is in turn relayed to the ventrobasal nuclei of the thalamus, and from there to layer IV of the cerebral cortex [78, 93].

The pattern of somatotopic organization of the main sensory nucleus and spinal nucleus of the trigeminal nerve is much the same as the topographical localization of oral areas in the thalamus and in the cerebral cortex. This suggests that the nerve fibers of the periodontal mechanoreceptors, which have their cell bodies in the trigeminal ganglion, by sending input to the main sensory nucleus and spinal nucleus of the trigeminal nerve give rise to subjective sensation (perception) of touch and pressure on the teeth, which is abolished by local alveolar anesthesia [36].

The periodontal mechanoreceptors that have their cell bodies in the mesencephalic nucleus of the trigeminal nerve are located inside the brain. There are strong projections from the mesencephalic nucleus to the motor nucleus of the trigeminal nerve that relay information from the periodontal mechanoreceptors to the jaw-closing muscles, which are involved in local (brainstem) reflexes during mastication. Similar reflexes at the segmental level may be relayed through periodontal fibers whose cell bodies are located in the trigeminal ganglion and project to the main sensory nucleus of the trigeminal nerve and in turn to the neurons of the trigeminal motor nucleus, resulting in an excitatory or inhibitory effect on the jaw-closing muscles. The periodontal mechanoreceptors that project to the trigeminal mesencephalic nucleus do not give rise to conscious perception of touch or pressure, since at present their cortical projections are ill-defined [93].

The periodontal ligament receptors signal only relatively small forces, and saturate when larger forces are applied. Larger forces are probably signaled by the mechanoreceptors in the bony sockets. The regulatory function of the mechanoreceptors of both the periodontal and bony socket in masticatory muscle contraction is important in orthodontic therapy. Specifically, the activation of the periodontal receptors, through mechanotransduction, may strengthen and/or normalize the forces of the masticatory muscles, which may be unsettling to the occlusion of the teeth. Two different periodontal reflexes have been described. First, brisk taps on a tooth elicit rapid and profound inhibition of activity in the jaw-closing muscles. This occurs, for example, when the teeth snap briskly together. The pathway for this reflex begins at the periodontal receptors, including those of the bony sockets, and travels up sensory trigeminal nerves to enter the brainstem mesencephalic nucleus of the trigeminal nerve and continues to trigeminal motor neurons. Here the signal is transformed by passing through one or two inhibitory interneurons, which then inhibit the motor neurons of the jaw-closing muscles. Inhibition of the motor neurons stops them from sending out action potentials to the jaw-closing muscles, which then cease to contract. This protective reflex response takes only about 0.01 s to act [78].

Periodontal receptors also contribute in a more complex manner to the control of the masticatory muscles. Pressing weakly on a tooth activates a different population of receptors in the periodontal ligament. When the signals from these receptors reach the trigeminal motor nucleus, the response is quite opposite to that caused by tapping on a tooth. Weak pressure induces a reflex excitation of the jaw-closing neurons which therefore increase the biting force. The function of this excitatory reflex is not entirely clear. When the teeth close onto a soft food bolus during chewing, the periodontal receptors are activated. The signals they send to the trigeminal motor nucleus induce a reflex activation of the jaw-closing muscles to make them contract a little more strongly to hold the bolus of food between the teeth, ready to be crushed. The same reflex is likely also to be important in guiding the teeth into occlusion. At the end of the chewing stroke, the occlusal surfaces of the teeth come together to crush the food bolus, but then grind the bolus even further by sliding across each other towards the occlusal position. This sliding, grinding action is the result of the activation of the appropriate masticatory muscles at the correct time when the teeth touch during chewing. A reflex emanating from the periodontal receptors is therefore well positioned to initiate and control this final grinding movement.

It is important that this grinding movement is made in the right direction, i.e., towards the occlusal position [78]. Because the periodontal receptors of the lower molars have a clear preference for the distal-lingual direction [93, 98], they may modulate the activity of the various jaw-closing muscles to ensure that the final grinding phase occurs in the correct direction, i.e., during the final phase of the chewing cycle, the lower molars on the working side approach the intercuspal position from a posterior and lateral position and thus they are likely to experience distal and lingually directed forces upon contact with the opposing upper molar teeth [93]. In this view the periodontal modulation of the masticatory muscles seems to be more complex. In other words, the periodontal receptors may be necessary to signal to the brain that the correct pattern of contact of opposing teeth has occurred before the reflex grinding movement to the final occlusal position is initiated. That is, if the cusps of the upper teeth do not meet the occlusal surfaces of the lower teeth in the correct relationship, the grinding movement is not initiated [78]. These recent neurophysiological studies may further suggest that in malocclusions of teeth the abnormal occlusal relationships of the teeth may not provide the appropriate periodontal sensory input to the cerebral cortex, resulting in sensorimotor integration dysfunction and hence, impairment of oral perception and of motor behavior [66, 71, 183].

8.15 Directional Sensitivity of Periodontal Receptors

Recent neurophysiological studies have indicated that the periodontal mechanoreceptors of the anterior teeth, which are more numerous than those of the posterior teeth, respond preferentially to horizontal forces. In contrast, the receptors supplying the molar teeth have a clear preference to distal-lingual direction of forces [93, 97, 98].

Accordingly, it has been suggested that since the periodontal mechanoreceptors of the anterior teeth clearly respond preferentially to horizontal forces, the rate of discharge of their nerves is highest when the stimulus is applied in the horizontal direction (most efficient excitatory stimulus of the periodontal receptors of the anterior teeth). These receptors are more numerous in the anterior segment of the dentition than in the posterior segment, attesting to the importance of a well-developed mechanoreceptive innervation in the anterior teeth that is involved in the initial stages of taking food into the mouth, and serves to guide the jaw into the intercuspal position, after the grinding movement of the teeth of the last phase of the chewing cycle has been completed [97, 98]. Use of the anterior teeth assures that only bite-size pieces of food are taken into the mouth. Sometimes the anterior teeth are even used as a “third hand” in manipulative tasks or as a precise cutting tool, suggesting heavy tactile and proprioceptive function of the anterior teeth [93].

The view that the periodontal receptors of the anterior teeth are involved in the guidance of the movement of the jaw to occlude in the intercuspal position might be consistent with the cybernetic model proposed by Petrovic and Stutzmann [158] of the physiological association between the occlusal relationship of the incisor teeth and the growth rate of the jaws. These investigators suggested that any deviation of the normal incisor relationship results in disruption of the physiological association leading to an abnormal relationship of the jaws.

8.16 Periodontal Receptors Regulate Tongue Posture. A Hypothesis

Neurophysiological studies have identified a distinct group of periodontal receptors, the "Ruffini-like ending" mechanoreceptors, with finger-like extensions in contact with the periodontal collagen fibers. These receptors are slowly adapting, meaning that they continue to respond (discharge) to sustained presence of a force level stimulus below about 1 N in magnitude (1 N corresponds to 98 g weight) in the anterior teeth, provided that the continuous force is applied in a horizontal direction. The corresponding value for the posterior teeth is higher, about 3–4 N. Above these limits most periodontal mechanoreceptors saturate and become progressively less sensitive to both the magnitude and the rate of increase of force on a tooth [93, 99, 155, 156].

Accordingly, it is hypothesized that the Ruffini-like ending mechanoreceptors of periodontal ligament have been programmed to regulate the tone of the tongue muscles through activation of hypoglossal motor nuclei. The tone of the tongue muscles in turn activates the discharge of the Ruffini-like ending periodontal mechanoreceptors by exerting a continuous horizontal pressure on the teeth. Thus, the Ruffini-like periodontal mechanoreceptors are actively involved in the regulation of tension of tongue muscles and hence of the posture of tongue in the oral cavity. When the tension of the tongue muscles increases above the 98 g weight force limit through central nervous system influences, the Ruffini-like ending periodontal mechanoreceptors saturate and stop providing the brain with useful information about the magnitude of a force applied to the teeth. This breakdown of central reference between the tone of the tongue muscles and the continuous discharge of the periodontal receptors may contribute to proprioceptive changes of the tongue position in the floor of the mouth, which in association with or without upper respiratory obstruction may lead to abnormal protracted tongue posture between the upper and lower anterior teeth, as occurs in the anterior open bite and tongue thrust malocclusion of teeth.

8.17 Heightened Perception of Tongue Position

Another neurophysiological detail to be considered about the tongue posture is that the different nerves that terminate in the tongue tip do not respond to tongue movement deformation, unless the receptive field of the sensory nerves at the tip of the tongue is brought into contact with the lower incisor teeth [93, 155, 156]. This is a multisensory integration function, associating sensory modalities coming from the same source (mouth), which increase the ability of the cerebral cortex to discriminate the position of the tongue in reference to oral structures, namely the anterior teeth, providing heightened oral perception of the walls of the oral cavity [15]. In this view, the position of the lower incisor teeth may be critical in determining the anterior boundary of the postural position of the tongue.

Lingual position–movement sensibility is jointly subserved by tactile and spindle endings. The tactile end organs are situated superficially in the mucosa and their discharge is activated by surface deformation occurring during tongue movement. The touch receptors signal the extent of tissue deformation that contraction of the lingual muscles produces. Lingual position–movement sensibility also involves the end organs situated deeply in the muscles, among them the muscle spindles. The distribution of spindle receptors within the tongue appears to be suited to recording length variations across different muscles and along the extent of a given muscle. Both the tactile and spindle receptors project into the same general loci in the cerebral somesthetic cortex. The region of the cerebral cortex in the precentral gyrus responsive to lingual spindle input in turn sends a large motor fiber projection to the hypoglossal nuclei. This projection is in part composed of large diameter fibers that terminate directly (monosynaptically) on the alpha motoneurons of hypoglossal nuclei which innervate the lingual musculature. The other part of the cerebral output is composed of gamma axons innervating the fusimotor–spindle system. The behavior of spindle receptors is classified in terms of tonic and phasic (possess both tonic and phasic) characteristics according to the frequency scale of the stimulus. Tonic receptors are depicted as being capable of signaling persistent stimuli, whereas phasic receptors are thought to comprise an "on-off" system signaling the application and release of a stimulus [214].

In this context, the tip of the tongue may be neurologically specialized to signal the anterior boundary of position–movement sensibility of the tongue through the "on-off" phasic spindle receptor system. When the tip of the tongue is in contact with the lower incisor teeth the spindle system is "on", signaling the anterior extent of the lingual postural position. This means that the tip of the tongue may be capable of accurately coding lingual spatial and temporal coordinates, and it is likely that the input of this information in the cortex is such that the latter is presented with a faithful central reflection of the differential pattern of lingual tip receptor excitation, whose ultimate expression is the articulation of a given sound in conjunction with a specific lingual target locus, most notably the anterior teeth. However, in aberrant spindle functioning or central dysfunction, such as neurotransmitter deficiency, the system may be shut down and the hypotonic tongue may occupy any position in the oral cavity (even a protrusive position between the incisors) suggesting a transient or permanent lingual neurological disorder (see Sect. 8.22). This view supports further the "equilibrium theory of tooth position" which is discussed in Sect. 8.19.

8.18 Mechanosensation in Bone

The receptors of the periodontal ligament and of the alveolar bony socket, the afferent nerves of muscles, tendons, and of the temporomandibular joints, are involved in the mechanosensory process. Specifically, Moss [91] and Moss and Cowin [176] have suggested that bone cells are connected by gap junctions permitting the intercellular transmission of ions and small molecules. In addition, gap junctions, as electrical synapses, permit electrical transmission in bone cells. All bone cells, except osteoclasts, are extensively interconnected by gap junctions to form an osseous connected cellular network (CCN), which underlies the organization of bone tissue. Each osteocyte, enclosed within its mineralized lacuna, has many cytoplasmic (canalicular) processes that interconnect with similar processes of neighboring cells. These processes lie within mineralized bone matrix channels (canaliculi). The small space between the cell process membrane and the canalicular wall is filled with macromolecular complexes.

Gap junctions are found where the plasma membranes of a pair of markedly overlapping canalicular processes meet. In compact bone, the canaliculi cross "cement lines" and they form extensive communications between osteons and interstitial regions. Gap junctions also connect superficial osteocytes to periosteal and endosteal osteoblasts. All osteoblasts are similarly interconnected laterally. Vertically, gap junctions connect periosteal osteoblasts with preosteoblastic cells, and these, in turn, are similarly interconnected. Effectively, each CCN is a true syncytium. Bone cells are electrically active. In a very real sense, bone tissue is "hard-wired."

The osseous CCN operationally appear to be analogous to hyperneuron function in the cerebral cortex, which is formed by many neurons connected by gap junctions (electrical synapses) permitting the synchronous firing of neurons in the hyperneuron. Hyperneurons in the central nervous system mediate conscious learning, memory perception and motor control of muscles (see Chap. 4).

Moss [91] has suggested that skeletal muscle contraction is a typical mechanotransduction process causing deformation (strain) of bone, and can initiate membrane action potentials in bone cells capable of transmission through interconnecting gap junctions to the entire osseous CCN. Bone organization is in turn "tuned" to precise frequencies of skeletal muscle activity. Periosteal osteoblasts may be directly stimulated by the contraction of muscles inserted into the periosteum. Anatomically, bone cells are competent mechanoreceptors. Their three-dimensional array of extensive canalicular cell processes is structurally well suited to sense deformation of the mineralized osseous matrix.

Accordingly, the oral experiential neural activity causing transformational changes in the cerebral cortex may also be involved in bone remodeling dynamics through the motor output frequencies to the masticatory muscles. In the "functional matrix" theory, the masticatory muscles are the periosteal matrices which are involved in bone morphogenesis [175]. The functional matrix approach to the growth and development of the oral–facial region is now widely accepted in orthodontic practice and teaching. Eminent researchers (for example, Bosma [67] and his coworkers) have suggested that oral postnatal growth and development is accomplished by progressive modifications of oral sensorimotor functions, which influence the growth of the skeleton and musculature of the mouth.

8.19 Equilibrium Theory of Tooth Position Reexamined

The fact that neurophysiological factors determine the tongue posture in the oral cavity (as described in Sects. 8.16 and 8.17) may also suggest that the normal arrangement of the teeth in the dental arch is essential for multisensory experiences and the activation of periodontal receptors through the tone of the tongue muscles. When the position of the teeth is altered the Ruffini-like periodontal mechanoreceptors may not be activated equally by the tongue pressure, and the tip of the tongue may not contact the lower incisor teeth, resulting in inappropriate input to the central nervous system and inadequate discriminative ability. Thus, the tongue–posture hypothesis raises again the old question of the position of the teeth being regulated by the balance of muscles inside and outside the mouth. To most orthodontists the equilibrium theory of tooth position means that the teeth occupy a position of stability between the opposing forces of the tongue on one side and the lips and cheeks on the other. At present most researchers consider simplistic the view that tooth position is a reflection of equilibrium between muscular forces acting inside and outside the mouth [198]. Recent neurophysiological studies, however, provide strong evidence which supports the assumption that the equilibrium theory may be valid.

Specifically, it has been suggested that the Ruffini-like ending periodontal mechanoreceptors and the Ruffini tactile receptors of the perioral skin, have similar neurophysiological properties, namely the presence of spontaneous activity, a steady discharge response to steady loads, relatively weak responses to the onset of stimuli, and directional sensitivity to loads [93, 159]. This means that the action potential frequencies of their nerves provide the same discriminative interpretation of their spatial and temporal elements in the cortex. These similarities of information input may also underlie similarities in motor responses to muscles with which the receptors are associated and by which they are activated. Specifically, the Ruffini-like ending periodontal mechanoreceptors have been assumed to be activated by the tone of the tongue muscles, while the Ruffini tactile receptors of perioral skin are activated by deformation of the skin, through contraction of the muscles of facial expression, for instance the orbicularis oris muscle [93, 155, 156].

Consequently, the tongue and the perioral muscles of facial expression may be involved actively in the balance of the position of the anterior teeth through the neurophysiological similarity of their respective receptors regulating the tension and the proprioceptive function of their respective muscles. Moss (see Sect. 8.18) has suggested that normal skeletal muscle activity produces intraosseous electric fields with morphogenetic influences in bone. Gap junctions between bone cells permit intercellular traffic of membrane action potentials, ions, proteins, and strands of DNA or RNA in the bone cellular network. Bone formation is "tuned" to precise frequencies of skeletal muscle activity. Accordingly since the frequencies in the tongue and perioral muscles have been assumed to be similar, it is reasonable also to assume that the bone organization inside and outside the mouth will be in balance with the frequencies of the tongue and perioral muscles, respectively.

8.20 Orthodontic Forces Regulate Alveolar Bone Remodeling Through Mechanosensation in Bone

It is interesting to note also that the sensitivity to stimulation only with horizontal forces of the Ruffini-like ending periodontal mechanoreceptors of the anterior teeth, including the first premolars, and of the Ruffini-tactile receptors of the perioral skin is about 25 g weight [93, 100]. This magnitude of force corresponds nicely with the orthodontic concept suggesting that most individual teeth respond favorably to biomechanical forces of at least of 25 g or 1 ounce (1 ounce = 28 g) [160]. This further suggests that the average orthodontic force that is capable of producing tooth movement through alveolar bone remodeling at the same time stimulates and elicits action potentials in the axon channels innervating the Ruffini-like ending periodontal receptors and the Ruffini-tactile perioral skin receptors. The action potentials of sensory nerves regulate the tension of the tongue and perioral muscles through central sensorimotor integration mechanisms. The motor output to the tongue and perioral muscles in turn mechanotransductively stimulates the sustained discharge of the Ruffini-like periodontal skin receptors and of the Ruffini-tactile perioral skin rcceptors. These activities in the tongue and perioral muscles are associated and proceed at the same time as the alveolar bone remodeling regulated by the frequencies of muscles, thus allowing the teeth to move to new positions within the limits laid down by the balance of the frequencies of the muscles (tongue and perioral) and of bone remodeling (see Sect. 8.19).

8.21 Forward Posturing of Tongue Tuned to Respiratory Rhythm: A Hypothesis

The respiratory center in the medulla is thought to act tonically, that is to have a continuous output of impulses to inspiratory muscles of the trunk and neck. Periodic inhibition of the discharge of the motor neurons to inspiratory muscles creates the respiratory rhythm, which can be modified by peripheral (sensory) and central inputs to motor neurons [190, 193].

The craniofacial muscles are involved in the adaptation of the oral cavity, as a portion of the upper respiratory tract, to the respiratory rhythm. Specifically, as described in the Sect. 8.19, the facial muscles (lips, nares) and the tongue muscles are activated by the continuous discharge of the tactile skin receptors and the Ruffini-like endings in the periodontal ligament, respectively. This might mean that the perioral and tongue muscles under certain circumstances are recruited with the frequency of the inspiratory muscles in the adaptation of the oral cavity to changes of respiratory rhythm. Similarly, the respiratory center can also drive the trigeminal jaw muscles in the same functional adaptation [191, 193]. These changes might in turn mean that the posture of the perioral, tongue, and jaw-opening muscles is tuned to the respiratory rhythm regulated by the afferent discharge.

Reflexes emanating from the laryngeal/pharyngeal sensory receptors and central influences to the motor neurons of the respiratory center may affect the rhythm of the inspiratory muscles, including the facial and tongue muscles [191]. The resulting changes in the contraction frequencies of the tongue muscles may affect the function of the Ruffini-like periodontal mechanoreceptors which saturate, meaning that they stop providing useful information to the brain as to the magnitude of the force exerted by the tongue pressure on the receptors located in the periodontal ligament. This breakdown of the central reference might result in muscle dysfunction, which may be associated with changes of tongue posture, such as tongue protrusion and jaw opening.

Changes in the respiratory rhythm and in the muscles may also be initiated following obstruction of the airway which thus increases the resistance of pulmonary ventilation. Accordingly, the inspiratory muscles must increase their work to overcome the resistance. The augmented effort in motor output is initiated by a reflex alteration (lower threshold) in the sensory receptors in the trunk and neck, including the receptors in the oral cavity. Thus, the tension of the involved muscles increases to assist the respiratory effort. Sessle [191] suggested that obstruction of the airway may cause the tongue to move forward through contraction of the posterior fibers of the genioglossus and hyoglossus muscles of both sides of the tongue. These muscles pull the base of the tongue forward such that the apex protrudes from the mouth. This action may be accompanied by opening the mouth and by associated facial and laryngeal muscle activities and inhibition of the antagonistic jaw-closing muscles [191].

The muscle activity protruding the tongue increases as the jaw is progressively opened. This effect is related to sensory information from the temporomandibular joint receptors. Thus, the forward posturing of the tongue produced by laryngeal/pharyngeal sensory input for airway maintenance may be enhanced by the activity of the temporomandibular joint receptors brought into action by the jaw-opening motion. These functions may be related to clinical observations of an association between increased lower face height and forward posturing of the tongue, and may be a factor in the etiology of anterior open bite malocclusion of the teeth [191, 193] through disruption of the muscular balance inside and outside the mouth.

Experimental studies in humans [191] and in monkeys [193] have indicated that the sensory information from a number of intraoral and perioral sites, including the teeth, can exert excitatory and/or inhibitory influences on the retrusive or protrusive motility of the tongue. The respiratory center is also involved in modifying tongue motility. The genioglossus muscles of the tongue demonstrate increased sustained discharge of the motor units suggesting hypertonicity of hypoglossal nuclei neurons, coupled with the tonicity of inspiratory muscles. Protrusion of the tongue is accompanied by contraction of the mylohyoid muscles which raise the floor of the mouth [170].

In sum, only by analysis of physiologically induced modifications in sensory feedback and/or central control to motor neurons can we begin to decipher which neurophysiological mechanisms and pathways become critical factors in altering neuromuscular function and the anterior posture of the tongue.

8.22 Tongue Thrust Activity

In clinical examinations in orthodontic and speech pathology, it has been observed that in some patients with malocclusion of the teeth the tongue tip protrudes against or between the anterior teeth during swallowing or speaking. This observation has led some clinicians to assign a causal relationship between thrusting of the tongue tip and the open bite of the teeth or excessive protrusion of the incisor teeth noted during examination. Tongue thrusting then in children and adults has produced a great deal of speculation and controversy as to the effects of this behavior on teeth position. Such a controversy is appropriate, as the remediation processes in myofunctional therapy for tongue thrusting involve a considerable investment of time, money, and patient motivation. While some orthodontists and speech pathologists feel strongly that tongue thrusting has a deleterious effect on tooth position, other clinicians contend that tongue thrusting is a normal transitional behavior and does not merit any therapy before puberty or, in most cases, ever [198].

The controversy about the possible effects of tongue thrusting behavior during swallowing on tooth position prompted the Joint Committee of the American Association of Speech and Hearing Association and the American Association of Orthodontists to issue the statement that "there is insufficient scientific evidence to permit differentiation between normal and abnormal or deviant patterns of deglutition, particularly as such patterns might relate to occlusion and speech. There is unsatisfactory evidence to support the belief that any patterns of movements defined as tongue thrust by any criteria suggested to date should be considered abnormal, detrimental, or representative of a syndrome" [161].

The recent neurophysiological studies, however, of Trulsson and Essick [93, 156] indicate that afferent nerves situated deep in the tongue muscles in the tip region (probably the terminal endings of muscle spindle afferents) discharge vigorously each time the subject protrudes the tongue. This means that the receptors located deep in the tongue muscles encode information of voluntary tongue movements. Voluntary movement, however, is conscious behavior associated with conscious perceptual discrimination of the boundaries of the oral cavity, which is closely related to neurocognition [15, 38]. Thus, the protrusion of the tongue might not be a simple involuntary (unconscious) reflex reaction of the genioglossus muscles pulling the base of the tongue forward into a protrusion position due for instance to dimensional restriction of the posterior airway. Instead protrusion of the tongue may be a consciously learned sensorimotor behavior developed in association with the learned oral sensorimotor function of speech, and representing a variation in the articulating speech patterns.

If this is the case, one of the key questions that then needs to be answered relates to whether the behavior of protruding the tongue could be dissociated from the speech pattern, or more appropriately, whether the tongue could be retrained during speech articulation. Myofunctional or speech therapy may prove helpful. Remember that the tongue muscles have motor units with a small number of muscle fibers. It follows that such muscles must be innervated by a large number of motor neurons. This enables the brain to grade the force of contraction of these muscles very accurately by progressively increasing the number of motor nerves that are activated, thereby giving many small increments of force. This is essential to enable the tongue to make extremely precise and rapid movements during speech.

Clearly, the exceptionally rapid tongue movements so evident in speech sound production could not take place if such movements had to proceed from a flaccid (hypotonic) muscle base, leading to speech articulatory dysfunction. Such hypotonicity of the tongue muscles may occur with aberrant spindle functioning or central dysfunction, such as neurotransmitters deficiency, resulting in severe depression of fusimotor spindle activity. The net effect is to depress the sensitivity of the lingual muscle spindles so that the activity in the alpha (motor) system is inappropriately related to changes in muscle length. Accordingly, the muscles of the tongue become flaccid or hypotonic due to absence of muscle tone. The decreased bias of the lingual fusimotor activity affects the articulation of speech, as well as the posture of tongue. The flaccid tongue occupies more space and is situated in a lower position in the floor of the oral cavity than the tongue with normal muscle tone. The lingual enlargement may in turn, cause obstruction of the airway which may predispose to protrusion of the tip of the tongue between the upper and lower incisor teeth both at rest and during speech. Thus the integrity of postural tone must be maintained in the tongue not only to keep the airway unobstructed but also to provide the necessary background upon which phasic lingual movements are superimposed [214]. Differential diagnosis of the enlarged hypotonic tongue (offering virtually no resistance to deformation or stretch) from macroglossia is needed to avoid unnecessary surgery. Present knowledge of myofunctional speech therapy indicates that by influencing the behavior of the spindle substrate with procedures such as lingual resistance exercises and motor training could provide a neural basis for conditioning motor output through centrally encoding motor responses to the fusimotor part of the muscle spindle via the gamma motor fibers [214].

On the other hand the swallowing pattern is considered innate or instinctive behavior, acquired in embryonic stages of development. Accordingly, the child may exhibit tongue thrusting during swallowing not as an adaptive behavior related to respiratory needs or the morphological characteristics of the mouth and pharynx, but as innate instinctive behavior, which usually changes spontaneously to adult patterns of tongue behavior, reflecting maturation of the central control mechanisms.

8.23 Motor Units in the Trigeminal System

A single alpha motor neuron in the trigeminal motor nucleus receives inputs from many sources (motor cortex, sensory receptors in the periodontal ligament, oral–facial region, muscle spindles of masticatory muscles, etc) and controls the activity of many skeletal muscle fibers. The number of muscle fibers controlled by one motor unit in different muscles can range from a fewer than ten up to many hundreds. It is important to note that all alpha motor neurons receive hundreds or even thousands of different inputs some of which are excitatory and others inhibitory. One motor neuron with the skeletal muscle fibers that it innervates is called a motor unit. Because each muscle fiber in a motor unit is innervated by the same motor neuron, it follows that one action potential in one motor neuron will generate one action potential simultaneously in all of the muscle fibers to which it is connected (neuromuscular junction). Consequently, all of the muscle fibers in a motor unit contract at the same time. This means, that the brain cannot control muscle fibers individually. Rather, it controls groups of muscle fibers that are combined into motor units. The number of muscle fibers per motor unit, therefore, determines the size of the force increments that can be used to grade muscle force, i.e., motor units with fewer muscle fibers such as in the tongue enable force to be graded more precisely [165, 188].

The number of muscle fibers in a motor unit varies with the function of the muscle. Muscles that carry out highly precise contractions, such as those in the tongue or those that move the eyes have motor units with a small number of muscle fibers. It follows that such muscles must be innervated by a large number of motor neurons. This enables the brain to grade the force of contraction of these muscles very accurately by progressively increasing the number of motor nerves that are activated, thereby giving many small increments of force. This is essential to enable the tongue to make extremely precise movements during speech. On the other hand in the less-precisely controlled muscles of the legs with more muscle fibers in each motor unit and relatively fewer motor neurons, the brain cannot produce finely graded force in these muscles. The size of the motor units in the masticatory muscles is rather in the middle of the range. That is, these muscles have motor units that are neither very small (such as the eyeball muscles) nor very large (such as the gastrocnemius). This middle size range is consistent with the masticatory muscle which requires good control, but less precision than is required for the movements of the eyes [165].

8.24 Classification of Motor Units

The classification of motor units as "fast" or "slow" is based on the rate at which they develop force during a single isometric twitch, i.e., the mechanical response to a single action potential occurring simultaneously in all of the muscle fibers in the motor unit by giving a single electric shock to its motor nerve. Under such laboratory conditions the fast-fatigable motor unit yields about 50 times more force than a single twitch in a type s (slow fiber). The precise values of the force produced by single isometric twitches in the three or perhaps four different types of motor unit in the human jaw muscles has not been established. However, it is likely that they are comparable with the twitch amplitudes (expressed in grams) of slow (s), fatigue-resistant (FR) and fast-fatigable (FF) motor units found in other muscles. Under normal circumstances, muscles contract when the nervous system sends continuous "trains" of action potentials down motor nerves to the motor units. These result in a smooth contraction of individual motor units and hence, of the whole muscle. The action potentials in different motor units are relatively asynchronous during voluntary contractions, which further smoothes the force generated by the whole muscle. The higher the frequency of action potentials in the motor unit, the more force it generates [165].

8.25 Activation of Motor Units in Muscle-Fiber Types

The fact that individual muscles contain different proportion of fiber types (it may well be that persons who eat unrefined foods have different proportions of fiber types in their jaw-closing muscles compared to people who eat modern soft diets and hence use their muscles less) raises the question of how the brain activates the right types of fibers or more accurately, the right motor units in any given muscle-fiber type. It has been suggested that during weak contractions the slow type fibers are always activated first, and exert force without fatiguing. Also, during weak contractions the blood flows through the muscle normally, so the delivery of oxygen and fuel can occur continuously [165].

However, as the brain calls for more and more force from the muscle, as for instance, when the teeth are clenched together with increasing force, the pressure inside the muscle increases. When the pressure reaches 20% of the maximal force that the muscle can produce, it begins to interfere with the flow of blood through the muscle. This reduction of perfusion of the muscle interferes with the performance of the slow fibers, which depend on a continuous supply of oxygen to maintain their aerobic metabolism. At about this stage the brain activates the fast (FR) fibers in addition to slow fibers. The aerobic and anaerobic capability of the FR fibers allows the muscle to continue to exert force even when the blood is limited in the muscle [165]. In a very powerful contraction, however, the brain recruits the FF fibers last of all, in addition to slow and FR fibers. In such muscular contraction no blood is entering the muscle. The FF fibers can continue to exert force because of their anaerobic metabolism, but their force falls quickly because of fatigue [165].

It has suggested that the superficial part of the temporalis muscle is rich in FF fibers in order to give acceleration and speed and, therefore, precision during function. The deeper posterior part of the same muscle is well endowed with slow fibers, which primarily have a postural role [168].

In malocclusions of the teeth with excessive overbite, which involves a combination of skeletal, dental and neuromuscular features that cause an undue amount of vertical overlap in the incisor region, there is a fast retraction of the mandible during closure through the FF superficial fibers of the temporalis muscle in order that the lower incisors avoid hitting the upper incisors, which are vertical and most often lingually inclined [1].

8.26 Plasticity of Muscles

Another important neurophysiological concept is that the metabolism of the skeletal muscle fiber is not fixed and immutable, but is a dynamic function, which may change throughout life to adapt to the functional demands made upon it [65, 166]. In this view the muscle fibers in motor units can be converted from one histochemical type to another by training. For example, increased use or endurance training based on lower intensity long-duration exercise can convert FF type II B muscle fibers to FR type II A fibers, which confers increased resistance to fatigue. This is because the FR fibers use a mixture of aerobic and anaerobic metabolism that enables them to exert moderate forces for quite prolonged periods, while at the same time are resistant to fatigue. The metabolic response of a muscle-fiber type to altered usage patterns is called plasticity of muscles [165].

Plasticity of muscles can also be produced in experimental animals. For example, experimental alteration of the oral sensory input to the central nervous system results in histochemical changes of the masticatory muscle-fiber type in rats, associated with changes in the growth of the jaws [167].

8.27 Fiber Type in Masticatory Muscles

Histochemical studies of human masseter and medial pterygoid muscles have indicated different proportions of muscle-fiber types termed types I, IIA, IIB, and IM. In fact the masticatory muscles have different populations of the various types of motor units in different parts of the muscles. This is known as compartmentalization. Type I fibers are relatively thin and they twitch slowly when stimulated using aerobic oxidation of ATP (these fibers have many mitochondria), which enables them to resist fatigue. The type IIA (FR) fibers twitch fast using a mixture of anaerobic and aerobic metabolism that enables them to exert moderate forces for quite prolonged periods when they contract and to have a moderate resistance to fatigue. In contrast the type IIB (FF) muscle fibers twitch fast using anaerobic conditions (glycolytic metabolism). The FF fibers contract extremely fast and exert very large forces. It is not surprising, therefore, that they fatigue very quickly as their energy is rapidly consumed [165]. The motor units with a faster contraction time can produce small or large amplitude twitches. The slow type motor units never produce large tension elements, yet the fast conduction axons can produce twitches that are both large and small [192].

Anaerobic metabolism is less efficient than aerobic metabolism. In the latter a net of 30 ATP molecules is formed per glucose molecule consumed, while during anaerobic conditions each glucose molecule consumed yields a net of only 6 ATP molecules [165].

The masticatory muscles differ in composition from limb or trunk muscles; each muscle is specialized in a manner that is adapted to its unique function depending on peripheral reflexes and the effects of central sensorimotor integration. For example, the lateral pterygoid muscle consists mainly of type I (slow) fibers, indicating a primarily postural stabilizing role on the mandible, in which large forces are not often required. In contrast the digastric muscle has many type II (FF and FR) fibers, which presumably confer the speed required for jaw movements during speech [165]. Type IM is the intermediate type that in the limbs is found only in muscles during development, exercise training, ageing, and in some pathological conditions. The masseter muscles of elderly human subjects have a reduced proportion of type I fibers and an increased proportion of type IM and type II fibers. In general the proportion of different types of fiber in the masticatory muscles appears to change with age, but in a manner that differs from one muscle to another, and that is different from what generally happens in limb muscles with age [165].

8.28 Hypertrophy of Masticatory Muscles

During clenching of the teeth in the intercuspal position the masticatory muscles contract isometrically without shortening. Increased muscle use or strength training based on high-force isometric contraction of the masticatory muscles may result in increased cross-sectional area of the individual muscle fibers and of the muscle as a whole. This leads to an increase in the maximal force that the muscle can exert, which is proportional to its total cross-sectional area. This phenomenon may correspond with the hypertrophic masseter and temporalis muscles that one sees in some patients who grind their teeth and have morphologically square faces [165].

In the hypertrophic masticatory muscles there is no increase in the number of muscle fibers, nor is there a change in the proportions of different fiber types. The overall blood supply of the muscle probably does not change, meaning that the same number of blood capillaries now supply a bigger muscle. There is, however, now good evidence that in the hypertrophic muscles much of the increased force that comes with strength training is the result of changes in the pattern of motor unit activation by the brain. This means that the increased force is due to functional neurological reprogramming of the central nervous system motor command to muscles as well as to structural changes in the muscles themselves [165].

8.29 Atrophy of Masticatory Muscles

Decreased use particularly with immobilization has given very variable results in experiments with different muscles, species, duration of immobilization, etc. Part of the apparent variability is no doubt due to the difficulty associated with controlling how much subjects contract muscles that have been immobilized by, for example, a plaster cast. It is usually reported that immobilization does not change the proportions of different fiber types, although different results have been obtained in studies of the masticatory muscles, as described below.

Immobilization of the mandible by intermaxillary fixation in monkeys has been reported to reduce both type I and type II fiber diameters by more than 50% in the masseter and temporalis muscles, although there was no change in the relative proportions of the two fiber types. In contrast, an increased proportion of type II fibers was found in the masseter muscles of patients 6–10 months after osteotomy to correct vertical maxillary excess overbite followed by 3–8 weeks of intermaxillary fixation [165].

Studies of altered activity patterns in masticatory muscles have not given consistent results in monkeys or humans. It has been claimed that in monkeys made edentulous for periods of up to 4 years, the mean cross-sectional area of type I fibers was reduced in the masseter and temporalis and there was overall a reduced oxidation capacity in the jaw muscles. In human subjects there was a reduced proportion of type II fibers with ill-fitting dentures and the cross-sectional area of individual fibers was also reduced [165].

These studies do not really lead to any firm conclusions on the effects on the composition of masticatory muscles from maneuvers that change the activity pattern of these muscles. Despite the lack of specific evidence it is fair to conclude that regular exercise by chewing foods of firm texture is likely to be important to maintain their function particularly in the elderly [165].

There are a range of pathological conditions (muscular dystrophies) that can rarely affect the masticatory muscles. Myasthenia gravis is an inherited muscle disease which illustrates how the neuromuscular junction works. There are receptors at the motor end plates of all skeletal muscle fibers that bind the neuromuscular transmitter acetylcholine and cause the motor end plate, and then the muscle, to depolarize. Patients with myasthenia gravis develop antibodies which destroy these receptors so that when the motor nerve releases acetylcholine, there are fewer sites to which it can bind. As a result the motor end plate and the muscle fiber membrane do not depolarize normally and hence will not contract. The result is muscle weakness and rapid onset of fatigue. Myasthenia gravis commonly affects the function of the facial muscles including those of the masticatory system, producing difficulties in chewing and swallowing. In children the most common form of muscular dystrophy is Duchenne's muscular dystrophy, which affects the muscles of the pelvis and limbs of young boys. The masticatory system can be involved later leading to malocclusions of the teeth [165].

8.30 The Hyperactive Child in Orthodontic Practice

This is a syndrome most obviously associated with psychosocial behavior with emotional overtones. It is considered a perceptual–motor disorder for it is hypothesized that its etiology stems from a disorder of the sensory system. The child reacts to stimulation of the touch–pressure receptors with a reaction of unpleasantness. He/she is apt to say "Ow, that hurts" when in your opinion there is insufficient basis for pain, or he/she will complain about your cold hands when other children do not. Also, the orthodontist often notices how easily some patients adapt to a new intraoral appliance and how difficult it is for others, so the new appliance itself is a sort of test of oral motor adaptability.

The hyperactive child reacts with a primitive emotional background of the nervous system. He/she knows that the presence of danger is often conveyed through the touch–pressure or temperature receptors. Survival is one of the fundamental laws of sensorimotor function, and the fundamental response to a possibly dangerous stimulus is fight or flight. Flight requires a locomotor response; fight requires emotional mobilization. The protective response of the tactile system of the organism or mouth is designed to cope with the fight or flight situation. The other senses—the discriminative system—play a lesser role in the phylogenetically older response of the tactile system. As the child matures, he learns cognitively to restrain the expression of negative feelings and to cooperate.

Under certain conditions, which are assumed to be associated with brain dysfunction, the balance of the protective and discriminative system is tilted in the direction of the protective system. The resulting centrifugal influences on receptor threshold and action potential transmission favor a nociceptive interpretation of tactile stimuli. The ARAS is implicated in alerting the brain to potential danger [4]. Fear produces adrenaline and adrenaline enhances readiness for fight or flight, but also lowers the threshold of tactile receptors.

In this context, there is a neurophysiological interaction between perception and emotions. This means that the nervous system geared to fight or flight is at the same time inhibiting the cortical process of perception and learning, resulting in depression of these abilities. This theoretical formulation provides some directions for a physiological basis of treatment of the hyperactive child. By simply shielding a child from distracting stimuli, we may provide the kind of sensory and emotional environment which will encourage a better balance between the protective and discriminative systems.

For example, it is suggested that we try touch–pressure stimuli, as opposed to light touch cutaneous stimulation, as a means of inhibiting a protective or nociceptive response to exteroceptive stimuli, applied especially to those parts of the body richly supplied with tactile receptors. These areas are the oral–facial region, the hands and the arms. We may also try the periodontal slow adapting mechanoreceptors, since these Ruffini-like endings, as discussed above in this chapter, have the same neurophysiological properties as the Ruffini tactile receptors of the oral–facial region.

Survival value is one of the strongest principles of sensorimotor function of the mouth. The protective system predominates in early life and often whenever development deviates from the normal or definite brain injury occurs. Predominance of the protective system inhibits manifestation of the discriminative system. Activation of the discriminative system inhibits the protective system. In other words, when one predominates, the other is inhibited. It is the adequate function of the discriminative tactile system that is required for sensory input-information processing, perception, learning, and motor planning in the brain. Thus, the conceptualization of a duality of function of the tactile system, provides guidance for more effective clinical procedures.

The deep limbic system (that includes the thalamic structures and hypothalamus, along with the immediate surrounding structures), especially the hypothalamus at the base of the brain, is responsible for translating our emotional state into physical feelings of relaxation or tension. The front half of the hypothalamus sends calming signals to the body through the parasympathetic nervous system. The back half of the hypothalamus sends stimulating or fear signals to the body through the sympathetic nervous system. The back half of the hypothalamus when stimulated is responsible for the fight or flight response, a primitive brain state that gets us ready to fight or flee when we are threatened or scared. This "hard-wired response" happens immediately upon activation by, for example, seeing or experiencing an emotional or physical threat. The deep limbic system is intimately connected to the prefrontal cortex and seems to act as a switching station between running an emotion (the deep limbic system) and rational thought and problem solving using our cerebral cortex. When the limbic system is turned on, emotions tend to take over. When it is cooled down, more activation is possible in the cortex. Current research shows a correlation between depression and increased deep limbic system activity and shutdown in the prefrontal cortex, especially on the left side [199].

8.31 Functional Jaw Orthopedic Appliances

At present it is difficult for the reader to gain a fair perspective about the proper place of functional appliance therapy in orthodontics and in teaching. Among the reasons of the disputes are how (or even if) such appliances really work. Functional jaw orthopedic appliances have only a theoretical basis. There is little neurophysiological evidence that functional appliances provide a more favorable environment for the developing dentition, the growth of the craniofacial skeleton and the function of the oral–facial musculature.

Functional appliances have been used on thousands of patients for many years and hundreds of articles have been written describing their utility, but there are differences of opinion about their modes of action. That they often succeed to the satisfaction of many discerning clinicians is not discussed—precisely how they work, why they do better in some cases than others, and how they achieve differential effects on muscle action, tooth development and bony changes are still not fully understood.

For example it has been suggested that a functional appliance causes contraction of the jaw-closing muscles during swallowing [177]. However, the same neuromuscular response is seen in children with tooth-together (normal) swallows [1]. From a neurophysiological point of view, if shortening of an elevator jaw muscle is arrested by the presence of a functional appliance with a bite plate between the teeth, gamma efferent activity will impose stretch upon the nuclear bag region of the intrafusal fibers. The result is increased facilitation of the alpha motoneurons of the trigeminal nucleus, resulting in stronger contraction of the jaw-closing muscle to overcome the resistance. If the stop is definitive, rapidly increasing tension of the isometric contraction will activate the tendon organs and cause inhibition of the jaw-closing muscle activity. During mastication this mechanism will cause unloading of the jaw-closing muscles due to the presence of a hard particle in the food or interference of a cusp not permitting displacement of the mandible [170]. Thus, the contention that the functional appliances create more favorable forces within the mouth which act against the teeth and bones but also are muscle-training devices, may not always be tenable.

In attempting to make clinical sense out of apparently conflicting research findings, it is well to remember that the central nervous system that regulates oral behavior is viewed as an impenetrable black box receiving inputs and emitting outputs (responses) where associations (habits) determine the quality and persistence of the motor behavior. While in the past these types of studies were popular, this was the extent to which researchers went in conceptualizing the nature of oral motor behavior. In Chap. 4 I introduce an approach to skill learning that examines the mental operations that intervene between perception and oral motor behaviors. The central tenet of this information processing approach is that in motor behavior there are a number of mental operations performed by the patient to solve a particular problem or perform a task. Information processing attempts to examine the role of the control center in sensing, attending to, transforming, retaining, and transmitting information. The appealing aspect of this approach is that it focuses attention on the cognitive activities that precede oral motor acts which are often neglected, and provides a rationale for linking cause and effect relationships during examinations. Cognition, by definition, involves awareness and judgment; i.e., cognitive activities imply "conscious awareness" of operations being performed.

The point is that an individual continually processes information from all modalities and actively operates on it to transform it for a particular use. Patients continuously register, process, and respond to a variety of stimuli such as pain, perturbations, movements, or speech sounds. With a little thought it is easy to realize that a great deal must go on between cognition and the execution of motor acts. Perhaps one of the most important roles for orthodontics in treating patients with motor disorders is to provide them with some insights about what is going on in their brain as they experience the functional jaw orthopedic therapy.

Clinical Brain Function 9

9.1 Introduction

The purpose of this chapter is to explain how the brain works, what happens when things go wrong, and how to optimize brain function. The chapter describes briefly the five brain systems that are most intimately involved with our behavior and make up much of what is uniquely human, as follows:

1. The limbic system lies at the center of the brain. It is the bonding and emotional mood control center. These fall very centrally into our concept of the psyche because they mediate the central emotional orientations that govern survival and our social interaction with others, including the capacity for love, hate, jealousy, etc. The limbic system being connected to others is essential to behavior, yet when this part of the brain is not healthy, people struggle with moodiness and negativity. Spending time with positively thinking people is essential to deep limbic health.
2. The prefrontal cortex at the front tip of the brain is the conscious supervisor, the part of the brain that helps to stay focused, make plans, controlling impulses, and making good (or bad) decisions.
3. The cingulate gyrus system is the part of the brain that allows the shifting of attention from thought to thought and between behaviors.
4. The system of temporal lobes underneath the temporal bones is involved with memory, understanding language, facial recognition, and temper control. Optimizing the function of this part of the brain helps us to experience inner peace.
5. The system of basal ganglia are large structures deep in the brain that surround the deep limbic system, controlling the body's idle speed. When this part of the brain works too hard (overactive), anxiety, panic, fearfulness, and conflict avoidance are often the result. Many of us may be vulnerable to anxiety and nervousness which may affect our relationships with patients. When the basal ganglia are underactive, people often struggle with concentration and fine motor control problems.

It is important to note that the brain systems are interconnected. When one system is affected, others are likely to be affected as well. Also, some brain researchers would separate the

M. Z. Pimenidis, *The Neurobiology of Orthodontics*,
DOI: 10.1007/978-3-642-00396-7_9, © Springer Verlag Berlin Heidelberg 2009

brain systems differently. For example, placing the cingulate system and deep temporal lobes within the limbic system. It is also interesting to note that many behavioral disorders formerly considered psychological actually have a biological basis documented with functional brain imaging [199].

9.2 Optimizing the Conscious Mind: The Prefrontal Cortex

Because the prefrontal cortex seeks stimulation for attention and awareness (arousal) many people unconsciously seek conflict as a way to stimulate their own prefrontal cortex. Without knowing (unconsciously) they try to upset you. They try to get you yell. They try to make you angry. It is important to have a calm demeanor in order to have a healthy prefrontal cortex function. Such people should do whatever they can not to yell or become emotionally intense. When they feel as if they are going to blow, they should take a deep breath or a break until they can get themselves under control. Those in whom the prefrontal cortex is not working hard enough, do not have access to good internal supervision, and they may say or do things impulsively. Thus, we should always think positive, and have healthy thoughts. Often people with prefrontal cortex problems have low self esteem and need encouragement and positive input from those they love. Disorganization is often a hallmark of prefrontal cortex dysfunction. Rather than complain about the disorganization, it is generally much more effective to help your partner become more organized [199].

The prefrontal cortex (especially the dorsolateral prefrontal cortex) is also the part of the brain that allows you to feel and express emotions; to feel happiness, sadness, joy, and love. This emotional environment is different from the primitive limbic system. Even though the limbic system controls mood and libido, the prefrontal cortex is able to translate the workings of the limbic system into cognizable feelings, emotions, and words such as love, passion, or hate. Underactivity or damage to this part of the brain often leads to a decreased ability to express thoughts and feelings. The prefrontal cortex has many connections to the limbic system. It sends inhibitory messages that keep it under control. The prefrontal cortex helps you "use your head along with your emotions." When there is damage or underactivity of this part of the brain, especially of the left side, the prefrontal cortex cannot appropriately inhibit the limbic system, leading to an increased vulnerability to depression if the limbic system becomes overactive. When people have prefrontal cortex problems, their emotions are often unavailable to them and their partners complain that they do not share their feelings [199].

9.3 Shifting Attention: The Cingulate Gyrus

The cingulate gyrus of the prefrontal cortex runs longitudinally through the middle deep aspects of the frontal lobes. The cingulate gyrus, often considered to be part of the limbic system, is covered separately. The cingulate gyrus is the part of the brain that allows you to shift your attention from one thing to another, to move from idea to idea, to see the options in life. Feelings of safety and security have also been attributed to this part of the brain. The term that best relates to this part of the brain is "cognitive flexibility."

Cognitive flexibility defines a person's ability to go with the flow, adapt to change, deal successfully with new problems. The cingulate system can be of a great help to starting a new job, learning a new system of doing things. Children in school need cognitive flexibility to shift learning from one subject to another. When the cingulate gyrus is working properly, we are more able to roll with the circumstances of the day. When it is impaired or overactive, cognitive flexibility is diminished. Along with shifting attention, cooperation is also influenced by the cingulate gyrus. People with cingulate problems have difficulty shifting attention and get stuck in ineffective behavior patterns. In other words people get stuck on unwanted, repetitive thoughts that they cannot get them out of their mind (obsessive–compulsive spectrum disorders, or OCD). In children OCD includes nail biting, pulling out their own hair, vocal tics, kleptomania, and eating disorders, such as anorexia nervosa and bulimia. The cingulate system has also been implicated in "seeing options" in planning and goal setting. In the profession of orthodontics, adaptable orthodontists readily utilize new ideas and techniques (after a scientific basis developed) and they are open to give their patients the latest information on what is new and exciting. Orthodontists who have cingulate gyrus problems tend to be rigid, do things the way they have always been done, and be autocratic ("Do it my way if you want me to treat you") [199].

9.4 Optimizing Limbic System Function

Lack of bonding and depression are often related. People who are depressed often do not feel like being around others and consequently isolate themselves, resulting in less bonding activity, a key substrate to limbic system dysfunction. Teenagers who are loved and connected with parents and teachers have a low incidence of pregnancy, drug use, and violence. Depression is known to be caused by a deficit of neurotransmitters, especially norepinephrine (noradrenaline) and serotonin. This deficit can cause increased metabolism or inflammation in the deep limbic system, which in turn causes many of the symptoms of depression. Antidepressant medications increase the availability of serotonin in the brain and they are often helpful to calm down cingulate hyperactivity, such as eating disorders, OCD, excessive worrying, etc. Low serotonin levels in the brain can be increased with foods high in carbohydrates, along with exercise. Carbohydrates contain the amino acid L-tryptophan, the precursor of the neurotransmitter serotonin, which has been found to be low in depressed patients. Physical exercise allows more of the L-tryptophan of the blood to enter the brain and be converted to serotonin in the neurons. Tryptophan is a relatively small amino acid, and it often has to compete with larger amino acids to cross the blood channels into the brain. With exercise the muscles of the body utilize the larger amino acids and decrease competition for tryptophan to enter the brain [199].

9.5
Optimizing Memory: The Temporal Lobes

Temporal lobe abnormalities occur much more frequently than previously recognized. Problems associated with abnormalities in either or both temporal lobes include [199]:

- Memory problems, amnesia
- Headaches or abnormal pain
- Anxiety or fear
- Abnormal sensory perceptions, such as visual or auditory distortions
- Aggression, directed internally or externally
- Dark or violent thoughts
- Emotional instability
- Dyslexia (reading difficulties)
- Illusions
- Moral or religious preoccupation
- Alzheimer's disease involving both temporal lobes

The following suggestions toward optimizing and healing the temporal lobes are given by Amen [199]:

- Singing has long been known to have healing qualities. Thus sing whenever and wherever you can. Singing will have a healing effect on your temporal lobes, and probably your limbic system as well. Actually all forms of vocalization, even simple talk can be therapeutic.
- Listen to a lot of beautiful music (music from country to jazz, from rock to classical). Music is one of the true joys of life and has healing properties. Scientific experiments have suggested that fetuses prefer Mozart and Vivaldi to other composers in early as well as later stages of pregnancy. Classical music can positively stimulate your brain. It has been suggested that complex music (classical) facilitates certain complex neuronal patterns involved in high brain activities such as mathematics and chess. By contrast, simple and repetitive music could have the opposite effect. It is likely that certain types of music open new pathways into the mind. Certain music may also be very destructive, for instance heavy metal music. Since the temporal lobes are involved in processing and producing rhythms, chanting, dancing, and other forms of rhythmic movement can be healing.
- Current research indicates that less than six hours sleep a night is associated with temporal lobes dysfunction, such as mood instability, decreased cognitive ability, and irritability.
- Clinical experience suggests that caffeine and nicotine are powerful vasoconstrictors that decrease blood flow to the brain, especially to the temporal lobes. Eliminate these substances from your body.
- Many people with aggressive behavior become much worse after a high sugar load. Instead a higher-protein/lower simple carbohydrate diet is very helpful.

9.6 Looking into Anxiety and Fine Motor Coordination: The Basal Ganglia

The basal ganglia are involved with integrating feelings, thoughts, and movement along with helping to shift and smooth motor behavior. In doing this the basal ganglia are involved in setting the body's "idle speed" or anxiety level. This is why we jump when we are excited, tremble when we are nervous, freeze when we are scared, or get tongue-tied.

Shifting and smoothing fine motor behavior is another function of the basal ganglia and is essential to hand writing and motor coordination. Enhancing the production of dopamine, the natural neurotransmitter of the basal ganglia, improves handwriting and overall motor coordination in children with attention deficit disorder. Parkinson's disease and Tourette's syndrome are caused by a deficiency of dopamine within the basal ganglia. People with an overactive function of the basal ganglia have a lot of energy and can become "movers in society." Using their energy effectively helps ward off anxiety.

Another interesting finding is that the basal ganglia are likely involved in the pleasure control loops of the brain. Brain imaging studies have indicated that cocaine is taken up mostly by the basal ganglia. Cocaine is a powerful enhancer of dopamine availability in the brain, and has both very fast uptake by the brain and very fast clearance from the brain. For this reason cocaine users get high doses, and when it is gone want more. This perpetuates the compulsive behavior and desire for the drug. It has been suggested that romantic love can have a cocaine-like effect on the brain, by robustly releasing dopamine in the basal ganglia. Love has real physical effects on the brain, as powerful as an addictive drug.

Excessive basal ganglia activity resets the body's idle to a revved-up level and can make people feel anxious, nervous, fearful, have tense muscles, and pessimistic. In such circumstances people do not have the physical or emotional energy to feel sexy or sexual, and they tend to shy away from sexual activity. Most of their memories are filled with anxiety and fear. They tend to wear out others by the constant fear they project [199].

9.7 Optimizing Neurotransmitter Deficiency: Nutritional Intervention

Over the past decade there has been significant research on food, nutrients, and depression. The results have surprised many people. We have been inundated by nutritional experts who tell us we should eat low-fat, high-carbohydrate diets. Unfortunately, low-fat is not the complete answer. Studies reported in the *American Journal of Psychiatry* indicate that men who are most likely to commit suicide have the lowest cholesterol blood levels. The deep limbic system needs fat in order to operate properly. Certainly some fats, for instance the omega-3 fatty acids found mostly in fish, are better than others. Dietary protein is also essential to a healthy deep limbic system. Proteins are the building blocks of the brain neurotransmitters. Low levels of dopamine, serotonin, and norepinephrine in the brain have all been implicated in depression and mood disorders. It is essential to eat enough protein in balanced amounts with fats and carbohydrates. Too much protein for some people may actually restrict the amount of amino acids that cross into the brain. Not enough protein will leave you with an amino acid deficit in the brain [199].

Low serotonin levels are often associated with worrying, moodiness, emotional rigidity, and irritability (a combination of deep limbic and cingulate gyrus problems). To enhance serotonin levels, eat balanced meals with carbohydrate snacks, such as whole-grain bread, yogurt etc. Low blood sugar often correlates with anger and irritability. Exercise can be of tremendous help along with nutritional supplementation.

Low norepinephrine and dopamine levels are often associated with depression, lethargy, trouble focusing, negativity, and mental fuzziness. To enhance norepinephrine and dopamine levels it is better to have protein snacks (meat, eggs, cheese) and to avoid simple carbohydrates, such as bread, pasta, cakes, and candy. Also supplements such as tyrosine and phenylalanine may be used in conjunction with advice from a doctor. A high-protein, low-carbohydrate diet that is relatively low in fat is recommended for children with attention deficit disorder. This diet has a stabilizing effect on blood sugar levels and helps with both energy and concentration. Unfortunately, the American diet is characterized by refined carbohydrates, which have a negative impact on dopamine levels in the brain and concentration.

Amen [199] has proposed the following diet to maintain high dopamine levels in the brain for attention and conscious awareness. He suggests that an ideal breakfast is an omelet with low-fat cheese and lean meat, such as chicken. An ideal lunch is a tuna, chicken, or fresh fish salad, with mixed vegetables. An ideal dinner contains more carbohydrates for balance with lean meat and vegetables. Eliminating simple sugars (such as cakes, candy, ice cream, pastries), and simple carbohydrates that are readily broken down to sugar (such as bread, pasta, rice, potatoes) will have a positive impact on energy level and cognition. It is important to note, however, that this diet is not ideal for people with cingulate gyrus or over-focus issues, which usually stem from a relative deficiency of serotonin. Serotonin and dopamine levels tend to counterbalance each other; whenever serotonin is raised, dopamine tends to be lowered and vice versa.

Orthodontic Avenues to Neuroplasticity 10

10.1 Introduction

The sensory and motor areas of the cerebral cortex are not rigidly hard-wired genetically. Neurogenesis and neurophysiology are dynamic. The allocation of a given region is a dynamic consequence of a series of interactive processes. The overall organization is not static, but derived from the dynamics itself. In visual development, for example, the retina and then the geniculate body and finally the cortex become organized, each deriving an organizing stimulus from the chaotic excitations established at the previous level. This cortical dynamic plasticity is preserved into later life, where injury, compensation, or a major new learned skill can result in development of new functional areas or in significant rearrangement of existing areas. For example, in a person studied before and after becoming a real-time translator at the United Nations, live positron emission tomography (PET) showed the development of a whole new language area [194].

Neurogenesis begins with embryogenesis, where neurons migrate up the glial cellular scaffold to make specific synapses. Neurogenesis is accompanied by growth and migration, and also by programmed cell death. Selective patterned cell death during development generates the architecture of the brain and its organization in maps that represent the external and internal worlds. This is readily appreciated by considering the sensory maps. For example, regions of the cerebral somatosensory cortex contain maps of the mouth plan, where the oral sensory modalities are anatomically represented in the precentral and postcentral gyri [2, 36].

Memory mechanisms have been built into the rules governing the architecture of the mouth in the gyri, so that activation of regions of the oral map by sensory experiences, such as for instance touch or pain, results in generation of action potentials in the neurons devoted to these modalities, resulting in sensory perception of the tactile or pain stimulus. The activation of the tactile or pain regions of the mouth in the somatosensory cortex is then relayed in the motor regions that plan and execute the movement of the mouth in response to oral sensory stimulation by experience. Thus, map structure leads to map function, which forms our reality. This implies that since the oral sensation and perception phenomena have a memory–learning component build into the brain through the experiences, the brain has a plasticity which is brought into play with changing degrees and types of oral sensory stimulation or sensory deprivation [2].

M. Z. Pimenidis, *The Neurobiology of Orthodontics*,
DOI: 10.1007/978-3-642-00396-7_10, © Springer Verlag Berlin Heidelberg 2009

This interrelationship between the cerebral cortical structure and function in response to changes in oral experiences is not considered in the classical theory and practice of orthodontics, of how the optimum of health, function, and esthetics, of the stomatognathic system may be obtained. Orthodontists may assume that only the dental alveolar structures can be altered by conventional orthodontic therapy, and may not suspect that the orthodontic changes, at least in the occlusion of the teeth, can modify the function and the development of the brain or structures within it.

10.2 Principal Components of Brain Function: The Neuron

The most rudimentary functional unit in the brain is the neuron, which typically has a cell body, a dendritic tree, and an axon as its output limb. The dendritic tree is the portion of the neuron where synaptic inputs are received. Some neurons receive tens of thousands of synaptic inputs distributed over their dendrites. These inputs may be either excitatory or inhibitory in nature; that is, they may result in a postsynaptic electrical potential that makes the dendritic cytoplasm either relatively less or more electronegative with respect to the extracellular space. Action potentials are properties of axons, while excitatory and inhibitory postsynaptic potentials occur at synapses. The excitatory and inhibitory postsynaptic potentials probably account for most electroencephalographic and magnetoencephalographic signals [151].

10.3 Bioenergetics

All the described activity requires energy, and it has been estimated that about 70% of the energy consumed by the brain is used for maintenance of the ionic gradients required for the initiation and propagation of postsynaptic potentials and action potentials. The remainder of the energy presumably is used for maintenance of the structural integrity of the cell and for various synaptic functions, including the synthesis and packaging of neurotransmitters [151].

The energy supplied to the brain is derived almost exclusively from glucose metabolism. The primary sources of brain glucose are glycogen stored in astrocytes and blood glucose. Some evidence suggests that glycogen within astrocytes is broken down to glucose and used within the astrocytes themselves, leaving neurons dependent on blood glucose for their energy needs. Blood glucose is transported across the blood–brain barrier by means of a saturable transport mechanism, the capacity of which can be modified to meet changes in local neural activity. Once within the cell, the glucose molecule is phosphorylated to glucose-6-phosphate by the enzyme hexokinase; this reaction is the critical point in the regulation of the rate of glycolysis. Glycolysis results in the net production of two adenosine triphosphate (ATP) molecules and a small amount of lactate. Usually, this bioenergetically inefficient glycolysis (anaerobic metabolism) is coupled with more efficient oxidative phosphorylation (aerobic metabolism). The final result is the production of between 30 and 35 ATP molecules per molecule of glucose, as well as the generation of substrates for the neurotransmitters acetylcholine, γ-aminobutyric acid, aspartate, and glutamate [151].

Interestingly and importantly, it has been suggested that with cortical activation, regional cerebral blood flow and glucose consumption increase out of proportion to blood oxygen extraction. In other words, regional cerebral blood flow and glycolysis are increased, but oxidative phosphorylation is not. Such decoupling of oxidative phosphorylation results in a relatively lower concentration of deoxyhemoglobin in the venous blood draining activated brain tissue [153].

10.4 Functional Brain Imaging in Orthodontics

Until very recently, scientists could only speculate about the brain's role in our personality and decision-making skills. We did not have advanced tools to look at the functioning of the brain and thus made many false assumptions about its impact on our lives. With the advent of sophisticated brain-imaging techniques, we are now answering questions about the brain's role in behavior that have practical applicability to our lives.

Functional brain imaging has gained enormous popularity in the radiological literature, as well as the basic and clinical neurosciences literature. This is because these powerful new in vivo imaging techniques focus on assessing the function of the brain, not just its structure. For example, functional magnetic resonance imaging (fMRI) is an entirely noninvasive technique to study many brain functions by contrasting the "resting" state with an "activated" state induced by a task designed to engage specific brain areas, including those involved in higher cognitive functions. Thus, fMRI techniques can be used to map cortical and subcortical areas subserving motor function in individual subjects, as well as dysfunctions and diseases of the motor system [151].

During growth and development the central nervous system guides the neurological development of the mouth. Conversely, the mouth is developmentally inscribed upon the maturing brain, and the neural networks of the brain are modulated by the oral sensory experiences. This ability of the brain for self-regulation of its structure according to oral sensory information (plasticity of the brain) is reflected in the anatomical representation of the mouth in the cerebral cortex [43, 65]. The structure and function of the cerebral cortex can be modified by oral sensory experiences, which regulate the input and processing of information from the senses in the brain through the electrical activity of nerves (action potentials) and the release of chemical signal molecules (neurotransmitters). Therefore the assumption is that the cerebral cortical representation of the mouth can become accessible to experimental observation following changes in the brain induced by the "stimulus" of orthodontic therapy.

In this view, when the neurons in a particular brain region are activated by specific cognitive tasks, both blood flow and the oxygen content of the blood in the region may increase. Oxygenated blood and deoxygenated blood have different magnetic properties. Neuronal activity in the brain region alters the local magnetic field by increasing the ratio of oxygenated to deoxygenated blood. The net result is a decrease in deoxyhemoglobin concentration in the blood. This change in the blood oxygenation levels in a particular region of the brain results in a signal that can be detected with the fMRI technique. The method has been termed the "blood oxygenation level-dependent," or BOLD, effect. Blood flow in the brain is believed to correspond closely to metabolic rate. Localized changes in blood flow are then taken to represent localized changes in brain metabolism induced by the functional maneuvers [151, 153] of orthodontic therapy. Methods such as PET and fMRI take advantage of these changes in local energy metabolism and hemodynamics to pinpoint increases in correlated neural activity [15].

fMRI, has a spacial resolution as good as 1 mm, and a temporal resolution of 1 s or less. fMRI affords the possibility of mapping cognitive and motor functions, in longitudinal studies, to very precise neuroanatomical structures, helping to identify structures and functional networks that may be abnormal. Images are normally acquired in the coronal plane (frontal), sagittal plane (dividing the body left and right) or axial plane (parallel with the long axis of the structure) [151, 153].

10.5 The Perceptual–Motor System of the Mouth

The organ referred to as the mouth is a major perceptual–motor system surrounded by a complex musculature. The sensory receptors are located in the oral mucosa, in the periodontal ligament of the mouth, in the bone, in the perioral skin, in the muscles and in the temporomandibular joints [154].

In the infant the motions of the mouth and tongue are controlled by the motor nuclei of the trigeminal nerve located in the pons and of the hypoglossal nerve in the medulla, respectively. In these nuclei the motor control of oral reflexes and of stereotyped functions, such as suckling, is modulated by the sensory input of the mouth. But the motions of the mouth are responsible for most of the sensory input that guides oral function. Thus, the mouth's sensory experiences are generated principally by its own actions, and its actions are responsive to its sensory experiences. In other words, in the oral area an intimate interaction occurs between the sensory and motor functions, especially an autocommunication [43]. This is the essential mechanism of the infantile suckle function and the background of biting and chewing functions. The mature patterns of mastication, however, appear gradually as new functions as a result of learning and memory, reflecting coincident maturation in oral sensory experiences and in the progressions of oral encephalization [195]. This establishment of the functions of the mouth in the cerebral cortex results in the highest achievements of sensorimotor integration function in the human body, which proceed apace with the changes in form and spatial relations of the tongue, lips, palate, teeth, and jaws during the growth and development period [43, 65].

However, the concurrently developing mouth and brain, intimately known at all stages of development, are liable to patterns of structural abnormality and dysfunction through impairment of cortical sensorimotor integration mechanisms which underlie sensation, perception, learning–memory, and motor control of muscles (see Sect. 3.3).

As we view the development of oral functions in this perspective, we are more aware that the mouth is essentially an information-processing system. Its various functions, such as speech, chewing, deglutition, taste, etc, are computational neural activities controlling the anatomy of the mouth and brain [91, 175, 196]. However, despite the available information about sensory receptors and afferent channels, and sensory neurophysiology, psychology and neurology, the integration of the sensory and motor functions of the mouth continue to evade our analysis.

10.6 Changes in Occlusion of the Teeth Change the Brain

The idea that changes at least in the occlusion of the teeth through orthodontic therapy may induce changes in the brain is an attractive hypothesis, which so far has not been investigated. There have been no studies in which orthodontic therapy would trained in part the brain hence improving the oral neuromuscular dysfunction by altering the maps in the motor cortex, where the mouth is represented very much out of proportion to its actual size, indicating the relative amounts of neural tissues devoted to motor functions of this organ. The remapping of the mouth in the cerebral cortex through orthodontic procedures is in accordance with the evidence that the cerebral cortex emerges postnatally under the partial influence of the sensory input and of information processing [102]. Thus the formation of a new oral representation map in the cerebral cortex may involve the growth and strengthening of the existing dendrite spinous synapses thus augmenting communication and sensory processing, the selective loss of spinous neural circuits and the formation of new ones, or most likely, combination of both processes [4–7].

10.7 Functional MRI Study of Malocclusions of the Teeth

Certain types of malocclusion of the teeth in children in which oral neurological development (sensorimotor dysfunction) and cerebral cortical development (sensorimotor integration dysfunction) are not fully correlated (designated as normal variation) can be studied with fMRI before and after orthodontic therapy. These exciting studies imply that the cortical neurons could learn new sensorimotor patterns through transformations of their cytoskeleton by the orthodontic experience, involving reconstruction of the occlusion pattern of the teeth and of the jaw's three-dimensional relationships.

The aim of such fMRI studies is to identify cortical and subcortical regions of the brain that show a different metabolic patterns in patients with malocclusion of the teeth before and after orthodontic therapy. We were able to carry out such a study in a 15-year-old boy with a severe class II division 1 with anterior open bite malocclusion associated with an abnormal motor tongue function (tongue thrust) during speech and swallowing, and impairment of oral sensory perception (as shown by oral stereognosis tests) without the use of vision. The preliminary results are exciting. It seems that after the orthodontic correction of the occlusion of the teeth at 17.5-years-old, most of the oral tasks, such as, speech and chewing movements, jaw and tongue movements (protrusive, retrusive, lateral) as well as tongue tip touching the palate movement, had activated cortical areas in both the left and right cerebral hemispheres controlling attention and awareness producing a correlated state with conscious perception. The activated areas included the prefrontal cortex, especially the dorsolateral prefrontal cortex, and areas controlling memory storage and language understanding in the temporal lobes [199]. These findings contrasted with the unilateral cortical activation of the same areas, mostly in the left hemisphere, before the orthodontic therapy. The findings might indicate that the learning potential of certain oral tasks was extended into both hemispheres, as indicated by the voluntary left and right coordinated chewing movements after correction of the occlusion of the teeth and jaw relationship.

Researchers have suggested that thalamocortical cells generating synchronous alpha and theta cortical activity in the electroencephalogram are linked by gap junctions in the thalamus [123], so that thalamocortical projections (or trans-corpus callosum pathways) could couple the two hemispheres in hyperneurons to account for bilateral firing synchrony [38]. Accordingly, one may assume that the unilateral prefrontal and temporal cortical activation before the orthodontic therapy underlay the dysfunction at least of the normal bilateral chewing function of the cerebral hemispheres (see Sect. 4.2).

The author is not a radiologist nor a clinical neuroscientist. In this study the resolution in the fMRI was 8 mm instead of 1 or 2 mm. Several other technical difficulties including imaging inexperience prevented proper performance and documentation of the research, and therefore its publication. However, since neuroimaging of oral neuromuscular behavior is new in the orthodontic literature, it may set the scene for future investigations in this exciting field.

11 Summary

Diagnosis of oral motor disorders is a professional skill. The basic goals of the practitioner are to discover the patient's motor problem, understand the potential causes of the problem, and make appropriate treatment recommendations. To accomplish these goals the orthodontic or dental clinician must have a basic knowledge of normal and disordered oral motor behavior. The central nervous system that regulates oral neuromuscular functions or behaviors is traditionally viewed as an impenetrable "black box". In this "black box" view the sensory and motor functions of the mouth are coded in the DNA molecule of the neuron, and the code acts as a blueprint or program containing all the necessary information for normal behavior development. This view has the corollary that evident oral neuromuscular disorders such as those seen in certain malocclusions of the teeth, have been interpreted as DNA coding errors, as mistakes in some inborn or congenital programs. In the past studies based on the "black box" view were popular, and this was the extent to which researchers went in conceptualizing the nature of oral motor behavior. Thus, most early oral motor behavior research contributed little toward the understanding of normal and dysfunctional behavior, since the intervening brain mechanisms, regulated by genetic factors, could not be influenced by orthodontic therapeutic measures.

However, the traditional dogma that the brain is a complex, but static, nongrowing, immutable structure, has been revolutionized by recent neuroscientific discoveries, which have indicated beyond any doubt that the brain is not simply governed by predetermined, immutable internal programs. Growth of the brain is also experience-driven. Mammalian brain emerges as an ever-changing, growing structure, remodeling autonomously and in response to experience. In other words, the brain is structured by the senses. Therefore, the oral sensory input and processing of information may cause correlated changes in the growth and development of the brain or in its component structures.

In this context, this book stresses a completely new conceptualization of neuromuscular functions or behavior of the mouth by examining the intervening brain mechanisms. The central tenet of this information input and processing approach is that in motor behavior there are a number of mental operations which are performed by the patient to solve a particular problem or perform a task. Information processing attempts to describe the functional and structural transformations of the cerebral cortex (neuroplasticity) caused by the sensory input which in turn underlie sensation–perception of oral experiences, and the learning and memory of experiences (cognitive functions), as well as motor output that controls the contraction of masticatory and speech muscles. The appealing aspect of this approach is that it

M. Z. Pimenidis, *The Neurobiology of Orthodontics*,
DOI: 10.1007/978-3-642-00396-7_11, © Springer Verlag Berlin Heidelberg 2009

focuses attention on the cognitive activities that precede oral motor acts, i.e., the motor acts have to be learned and inscribed in memory before the motor command is issued to the muscles. This provides a rationale for linking cause and effect relationships. The point is that an individual continually processes sensory information from all modalities and actively operates on it to transform it for a particular use. The oral information-processing model offers a way of understanding the improvement in neuromuscular dysfunction following the completion of orthodontic treatment.

Orthodontic procedures stimulate the oral–facial mechanosensory systems so that afferent nerves carrying information (action potentials) from the periodontal, periosteal, muscle, tendon, skin, and oral mucosa receptors converge, through sensory pathways, to dendritic spinous synapses of the corresponding neurons of the oral–facial somatosensory cortex, where the information is pooled and integrated in the cytoskeletal microtubules of the subsynaptic zone lying beneath the spinous synapses. The microtubules in the subsynaptic zone of the dendrites are the physical substrate for the computation and storage of memory of learned experiences. The resulting motor output controls the contraction of the oral–facial muscles. Bone remodeling is "tuned" to the precise frequencies of skeletal muscle activity dictated by the output of the motor cortex. Bone cells are activated by contraction of the muscles mechanotransductively, and can initiate membrane action potentials capable of transmission throughout the entire bone cellular network, via interconnecting gap junctions (electrical synapses). Thus, the information processing approach may be used to describe the contents of the formerly impenetrable "black box," such as viewing the transformation of oral information as a continuum of brain changes linked by a series of complex researchable mental processes, such as perception, memory, motor response, feedback, and bone reorganization.

The representation of the mouth in the sensory and motor maps in the postcentral gyrus and precentral cortical gyrus, respectively, is not rigidly hard-wired. The overall cortical organization is also experience-driven. This implies that the neural events that are correlated with learning and memory of oral experiences are incorporated into the rules that govern the architecture of the oral cortical maps. This further implies that the oral neuromuscular dysfunction associated with malocclusion of the teeth could be improved through the transformation of the corresponding sensory and motor map areas of the cerebral cortex, induced by the orthodontic therapy. This insight is emancipating. Orthodontic neuroplasticity enables the practitioner to view the patient as an active processor of sensory stimuli provided by the orthodontic procedures with an impact right into the heart of the neuromuscular dysfunction, the brain.

Orthodontic education has been pretty much conservative by claiming that only the dental alveolar structures can be clearly changed by orthodontic treatment of malocclusions of the teeth. The enormous representation of the oral structures in the cerebral cortex is not taken into account. Classical orthodontic theory also does not take into account the fact that the concurrently developing mouth and brain (intimately known at all stages of development) can show prenatal and postnatal patterns of abnormalities, and that such abnormalities affect the learning ability of the brain and the maturation of oral sensorimotor functions, which according to the "functional matrix" theory [175, 196] underlie the skeletal structure of the mouth. How the optimum neuromuscular function, stability of the occlusion of the teeth, health and esthetics of the stomatognathic system may be obtained through neuroplasticity is not considered in classical orthodontic theory and practice, and it may be the time for orthodontic education to move in these new directions.

References

1. Moyers RE (1988). Handbook of Orthodontics, 4th edition. Year Book Medical Publications, Chicago, IL.
2. Black IB (2002). The Changing Brain. Oxford University Press, New York.
3. Priel A, Tuszynski JA and Cantiello HF (2006). In: Tuszynski JA (ed), The Emerging Physics of Consciousness. Springer, Berlin, pp. 223–325.
4. Yuste R, Gutnick MJ, Saar D, Delaney KR and Tank D (1994). Neuron 13:23–43.
5. Hering H and Sheng M (2001). Nat. Rev. Neurosci. 2:880–888.
6. Matus A (2000). Science 290:754–758.
7. Shepherd GM, Brayton MJP, Segev I, Rinzel J and Rall W (1985). Proc. Natl. Acad. Sci. USA 82:2192–2195.
8. Engert F and Bonhoeffer T (1999). Nature 399:66–70.
9. Toni N, Bucks PA, Nikonenko I, Bron CR and Muller D (1999). Nature 402: 421–425.
10. Fifkova E and Delay RJ (1982). J. Cell. Biol. 95:345–350.
11. Fisher M, Kaech S, Knutti D and Matus A (1998). Neuron 20:847–854.
12. Scott EK and Luo L (2001). Nat. Neurosci. 4:359–365.
13. Schroeder CE and Foxe JJ (2002). Brain Res. Cogn. Brain Res. 14(1):187–198.
14. Seitz A, Kojima H, Oiwa K, Mandelkow EM, Song VH and Mandelknow E (2002). EMBO J. 21(18):4896–4905.
15. Woolf NJ (2006). In: Tuszynski JA (ed), The Emerging Physics of Consciousness. Springer, Berlin, pp. 49–94.
16. Hebb PO (1949). The Organization of Behavior. Wiley, New York.
17. Bliss TVP and Lomo T (1973). J. Physiol. Lond. 232:331–356.
18. Morris RGM, Anderson E, Lynch GS and Baudry M (1986). Nature 319:774–776.
19. Barde VA (1989). Neuron 2:1525–1534.
20. Chao MV (1992). Neuron 9:583–593.
21. Lo DC (1995). Neuron 15:971–981.
22. Black IB (1999). J. Neurobiol. 41:108–118.
23. Goldman SA (1998). J. Neurobiol. 36:276–286.
24. Kirschenbaum B, Nedergaard M, Preuss M, Barami K, Fraser RAR and Goldman SA (1994). Cereb. Cortex 4:576–589.

25. Gnahn H, Hefti F, Heumann ME, Schwab ME and Thoenen H (1983). Dev. Br. Res. 9:45–52.
26. HeftiHeftier F, Dravid A and Hartikka J (1984). J. Br. Res. 293:305–311.
27. Martinez HJ, Dreufus CF, Jonakait GM and Black IB (1987). Br. Res. 412:295–301.
28. Johnson EM Jr, Rich KM and Yip HK (1986). Trends Neurosci. 9:33–37.
29. Martin DP, Ito A, Horigome K, Lampe PA and Johnson EM Jr (1992). J. Neurobiol. 23:1205–1220.
30. Leibrock J, Lottspeich F, Hohn A, Hofer M, Hengerer B, Masaiakowski P, Thoenen H and Barde YA (1989). Nature 341:149–152.
31. Greene LA and Shooter EM (1980). Ann. Rev. Neurosci. 3:353–402.
32. Greene LA and Tischler LA (1976). Proc. Natl. Acad. Sci. USA 73:2424–2428.
33. Tuszynski JA and Woolf N (2006). In: Tuszynski JA (ed), The Emerging Physics of Consciousness. Springer, Berlin, pp. 1–26.
34. Dai Z, Lercher LD, Qu P, Wu H, Black IB and Dreyfus C (1996). Soc. Neurosci. 22.
35. Tanzi RE and Parson AB (2000). Decoding Darkness. The Search for the Genetic Causes of Alzheimer's Disease. Perseus, Cambridge, MA.
36. Noback CR (1967). The Human Nervous System. McGraw-Hill Book, New York.
37. Popper KP and Eccles J (1983). The Self and its Brain. Routledge and Kegan Paul, London.
38. Hameroff S (2006). In: Tuszynski JA (ed), The Emerging Physics of Consciousness. Springer, Berlin, pp. 193–253.
39. Yuste R and Bonhoeffer T (2001). Ann. Rev. Neurosci. 24:1071–1089.
40. Lenner B, Falduto J and Shors TJ (2003). J. Neurosci. 23(2):659–665.
41. Dehmelt L, Smart FM, Ozer RS and Halpain S (2003). J. Neurosci. 23(29): 9479–9490.
42. Spielman AI and Ship JA (2004). In: Miles TS, Nauntofte B and Svensson P (eds), Clinical Oral Physiology. Quintessence, Copenhagen, pp. 53–70.
43. Bosma JF (1970). Second Symposium on Oral Sensation and Perception. Charles C. Thomas, Springfield, IL, pp. 550–555.
44. Granit R (1962). Receptors and Sensory Perception. Yale University Press, New Haven.
45. Greenberger RS (1966). A study of pressoreceptive thresholds of teeth in children. Master's Thesis, Univ. of Michigan, Ann Arbor.
46. Mel BW (1999). In: Stuart G, Spruston N and Hauser M (eds), Dendrites. Oxford University Press, New York.
47. Jaslove S (1992). Neuroscience 47:495–499.
48. Johnston D, Magel JC, Colbert CM and Christie BR (1996). Ann. Rev. Neurosci. 19:165–186.
49. Rihn LL and Claiborne B (1990). Dev. Brain Res. 54:115–124.
50. Greenough W (1975). Am. Sci. 63:37–46.
51. Kaech S, Parmar H, Roelandse M, Bornmann C and Matus A (2001). Proc. Natl. Acad. Sci. USA 98:7086–7092.
52. Matus A, Ackermann M, Pehling G, Byers HR and Fujiwara K (1982). Proc. Natl. Acad. Sci. USA 79:7590–7594.
53. Aristotle (1966). Tredennick H (ed), Posterior Analytics book IIxix. Loeb Classical Library. Harvard University Press, pp. 257–258.

54. Khuchua Z, Wozniak DF, Bardgett ME, Yul Z, McDonald M, Boero J, Hartman RE, Sims H and Strauss AW (2003). Neuroscience 119(1):101–111.
55. Woolf NJ (1998). Prog. Neurobiol. 55:59–77.
56. Woolf NJ (1996). Neuroscience 74(3):625–651.
57. Woolf NJ and Hameroff SR (2001). Trends Cogn. Sci. 5:472–478.
58. Davia CJ (2006). In: Tuszynski JA (ed), The Emerging Physics of Consciousness. Springer, Berlin, pp. 255–292.
59. Skaggs WE and McNaughton BL (1996). Science 271:1870–1873.
60. Shors TJ and Dryver E (1994). Br. Res. 666:232–238.
61. Shors TJ and Servatius RJ (1995). Neuroreport 6:677–680.
62. Mountcastle VB (1975). Johns Hopkins Med. J. 136:109–131.
63. Kunbzansky PE and Leiderman PH (1971). In: Solomon P, Kubzansky PE, Leiderman PH, Mendelson JH, Trumbull R and Wexler D (eds), Sensory Deprivation. A Symposium. Harvard University Press, Cambridge, MA, pp. 221–238.
64. Lindsley D (1971). In: Solomon P, Kubzansky PE, Leiderman PH, Mendelson JH, Trumbull R and Wexler D (eds), Sensory Deprivation. A Symposium. Harvard University Press, Cambridge, MA, pp. 174–194.
65. Bosma JF (1967). Symposium on Oral Sensation and Perception. Charles C. Thomas, Springfield, IL, pp. 350–352.
66. Kawamura Y (1972). In: Emmelin N and Zotterman Y (eds), Oral Physiology. Pergamon Press, Oxford, pp. 183–204.
67. Bosma JF (1967). Symposium on Oral Sensation and Perception. Charles C. Thomas, Springfield, IL, pp. 98–110.
68. Sechenov I (1970). In: Gibbons G (ed), Reflexes of the Brain. The MIT Press, Cambridge, MA.
69. Bruner JS (1971). In: Solomon P, Kubzansky PE, Leiderman PH, Mendelson JH, Trumbull R and Wexler D (eds), Sensory Deprivation. A Symposium. Harvard University Press, Cambridge, MA, pp. 195–207.
70. Field TM (1995). Touch in Early Development. Lawrence Erlbaum, Mahwah, NJ.
71. Kubie LS (1971). In: Solomon P, Kubzansky PE, Leiderman PH, Mendelson JH, Trumbull R and Wexler D (eds), Sensory Deprivation. A Symposium. Harvard University Press, Cambridge, MA, pp. 58–71.
72. Solomon P , Kubzansky PE, Leiderman PH, Mendelson JH, Trumbull R and Wexler D (eds) (1971). Sensory Deprivation. A Symposium. Harvard University Press, Cambridge, MA, pp. 1–5.
73. Schanberg S (1995). In: Field TM (ed), Touch in Early Development. Lawrence Erlbaum, Mahwah, NJ, pp. 67–79.
74. Naito E and Sadato N (2003). Rev. Neurosci. 14(4):387–399.
75. Illingworth RS (1991). The Normal Child, 10th edition. Churchill Livingstone, Edinburgh.
76. Dubner R, Sessle BJ and Storey AT (1978). The Neural Basis of Oral and Facial Function. Plenum, New York.
77. Békésy Gvon (1971). In: Dubner R and Kawamura Y (eds), Oral-Facial Sensory and Motor Mechanisms. Appleton-Century-Crofts, New York, pp. 1–6.
78. Miles TS (2004). In: Miles TS, Nauntofle B and Svensson P (eds), Clinical Oral Physiology. Quintessence, Copenhagen, pp. 219–243.

79. Emmons PG and Andersoon L (2005). Understanding Sensory Dysfunction. Jessica Kingsley, London.
80. Levy DM, Quoted by Kubie LS (1971). In: Solomon P, Kubzansky PE, Leiderman PH, Mendelson JH, Trumbull R and Wexler D (eds), Sensory Deprivation. A Symposium. Harvard University Press, Cambridge, MA.
81. Heron WH (1971). In: Solomon P, Kubzansky PE, Leiderman PH, Mendelson JH, Trumbull R and Wexler D (eds), Sensory Deprivation. A Symposium. Harvard University Press, Cambridge, MA, pp. 6–33.
82. Freedman SJ, Grunebaum HU and Greenblat M (1971). In: Solomon P, Kubzansky PE, Leiderman PH, Mendelson JH, Trumbull R and Wexler D (eds), Sensory Deprivation. A Symposium. Harvard University Press, Cambridge, MA, pp. 58–71.
83. Ayer WA (1979). In: Oral Motor Behavior. U.S.A. DHEW Publ. No (NIH) 79-1845, pp. 7–22.
84. Baer DM (1979). In: Oral Motor Behavior. U.S.A. DHEW Publ. No (NIH) 79-1845, pp. 28–31.
85. McFadden J (2006). In: Tuszynski JA (ed), The Emerging Physics Of Consciousness. Springer, Berlin, pp. 387–406.
86. Ruff GE, Levy EZ and Thaler VH (1971). In: Solomon P, Kubzansky PE, Leiderman PH, Mendelson JH, Trumbull R and Wexler D (eds), Sensory Deprivation. A Symposium. Harvard University Press, Cambridge, MA, pp. 72–90.
87. Pimenidis MZ (2001). The Internat. J. MEAW Technic Res. Found. 8(1):10–17.
88. Morimoto T and Kawamura Y (1978). In: Perryman JH (ed), Oral Physiology and occlusion. An International Symposium. Pergamon Press, New York, pp. 149–169.
89. Teuber HL, Krieger HP and Bender MB (1949). Fed. Proc. 8:1958.
90. Haber WB (1955). J. Psychol. 40:115.
91. Moss ML (1997). Am. J. Orthod. Dentofacial Orthop. 112:221–226.
92. Swazey JP (1969). Reflexes and Motor Integration: Sherrington's Concept of Integrative Action. Harvard University Press, Cambridge, MA.
93. Trulsson M and Essick GK (2004). In: Miles TS, Nauntofte B and Svensson P (eds), Clinical Oral Physiology. Quintessence, Copenhagen, pp. 165–197.
94. Slavicek R (2001). Int. J. MEAW Technic Res. Found. 8(1):18–40.
95. Yemm R (1979). U.S.A DHEW No (NIH) 79–1845, pp. 138–156.
96. Thomson H (1971). Br. Dent. J. 130:187–193.
97. Trulsson M, Johansson RS and Olsson KA (1992). J. Physiol. 447:373–389.
98. Johnsen SE and Trullson M (2002). J. Neurophysiol. 89:1478–1487.
99. Trulsson M and Johansson RS (1994). J. Neurophysiol. 72:1734–1744.
100. Trulsson M (1993). J. Neurophysiol. 69:474–481.
101. Ayres AJ (1979). Sensory Integration and the Child. Western Psychological Services, Los Angeles.
102. Johnson MH (1997). Developmental Cognitive Neuroscience, Chapter 6. Blackwell, Oxford.
103. Woolf NJ (1993). J. Chem. Neuroanat. 6(6):375–390.
104. Woolf NJ, Zinnerman MD and Johnson GV (1999). Brain Res. 821(1):241–249.
105. Gronly-Dillon J and Perry GW (1979). J. Physiol. 293:469–484.
106. Woolf NJ (1997). Conscious. Cogn. 6:574–596.

107. Woolf NJ (1999). Trends Neurosci. 22:540–541.
108. Gray JA (1995). Behav. Brain Sci. 18:659–722.
109. Dayhoff J, Hameroff S, Lahoz-Beltra R and Swenberg CE (1994). Eur. Biophys. J. 23:79–83.
110. Hausser M, Spruston N and Stuart GJ (2000). Science 290:739–750.
111. Gundersen G (1997). Biomed. Front. 4, Special Section.
112. Bohm D and Hiley BJ (1993). The Undivided Universe. Routledge, New York.
113. Penrose R and Hamerroff SR (1995). J. Conscious. Stud. 2:98–112.
114. Traub RD, Kopeli N, Bibbig A, Buhl EH, Le Beau FE and Whittington MA (2001). J. Neurosci. 21(23):9478–9486.
115. Traub RD, Draguhn A, Whittington MA, Baldeweg T, Bibbig A, Buhl EH and Schmitz D (2002). Rev. Neurosci. 13(1):1–30.
116. Kandel ER, Schwartz JS and Jessell TM (2000). Principles of Neural Science, 4th edition. McGraw-Hill, New York.
117. Gray CM and Singer W (1988). Proc. Natl. Acad. Sci. USA 86:1698–1702.
118. Gray CM, Koning P, Engel AK and Singer W (1989). Nature 338:334–337.
119. Varela FJ (1995). Biol. Res. 28:81–95.
120. Hatton GI (1998). Cell Biol. Int. 22:765–780.
121. Froes MM, Correia AH, Garcia-Abreu J, Spray DC, Campos de Carvalho AC and Neto MV (1999). Proc. Natl. Acad. Sci. USA 96(13):7541–7546.
122. VenancePenances L, Rozov A, Blatow M, Burnashev N, Feldmeyer D and Monyer H (2000). Proc. Natl. Acad. Sci. USA 97(18):10260–10265.
123. Hughes SW, Lorincz M, Cope DW, Blethyn KL, Kekesi KA, Parri HR, Juhasz G and Cruneli V (2004). Neuron 42(2):253–268.
124. Troelstrup B and Möller E (1970). Scand. J. Dent. Res. 78:425–430.
125. Möller E (1975). In: Rowe NH (ed), Occlusion: Research in Form and Function. A Symposium. University of Michigan School of Dentistry, The Dental Research Institute, pp. 57–89.
126. Flohr H (1995). Behav. Brain Res. 71:157–161.
127. Flohr H (2000). In: Metzinger T (ed), Neural Correlates of Consciousness. Empirical and Conceptual Questions. MIT Press, Cambridge, MA, pp. 245–258.
128. Bosma JF (1973). Fourth symposium on Oral Sensation and Perception. U.S.A DHEW Publ. No (NIH) 73-546, Bethesda, Meryland, pp. 3–8.
129. Sessle BJ (1979). In: Oral Motor Behavior. U.S.A. DHEW Publ. No (NIH) 79-1845, pp. 128–133.
130. Bishop B (1979). In: Oral Motor Behavior. U.S.A. DHEW Publ No (NIH) 79-1845, pp. 83.
131. Goldberg L (1979). In: Oral Motor Behavior. U.S.A. DHEW Publ No (NIH) 79–1845, pp. 83.
132. Battin RR and Haug CO (1968). Speech and Language Delay, 2nd edition. Charles C. Thomas, Springfield, IL.
133. Bierman DJ and Whitmarsh S (2006). In: Tuszynski JA (ed), The Emerging Physics of Consciousness. Springer, Berlin, pp. 27–48.
134. Saito DN, Okada T, Morita Y, Yonekura Y and Sadato N (2003). Brain Res. Cogn. Brain Res. 17(1):14–25.

135. Fu KM, Johnston TA, Shah AS, Arnold L, Smiley J, Hackett TA, Garraghty PE and Schroeder CE (2003). J. Neurosci. 23(20):7510–7515.
136. Taylor-Clarke M, Kenneth S and Haggard P (2002). Curr. Biol. 12(3):233–236.
137. Halpain S (2000). Trends Neurosci. 23(4):141–146.
138. Libet B (2004). Mind Time. The Temporal Factor in Consciousness. Harvard University Press, Cambridge, MA.
139. Gray JA (2004). Consciousness: Creeping up on the Hard Problem. Oxford University Press, Oxford.
140. Watson WE (1976). Cell Biology of Brain. Chapman and Hall, London, pp. 160–177.
141. Gibson JJ (1979). The Ecological Approach to Visual Perception. Houghton Mifflin, Boston.
142. Medawar P and Medawar J (1985). Aristotle to 2005. Oxford University Press, pp. 1 and 235.
143. Kreiter AK and Singer W (1996). J. Neurosci. 16:2381–2396.
144. Singer W (1999). Neuron 24:49–55.
145. Perry HT, Lammie GA, Main J and Teuscher GW (1960). J. Am. Dent. Assoc. 60:626–633.
146. Yemm R (1971). Arch. Oral Biol. 16:269–273.
147. Paulesu E, Frith CD and Frackowiak RS (1993). Nature 362:342–345.
148. Bressman T (2004). In: Miles TS, Nauntofte B and Svensson P (eds), Clinical Oral Physiology. Quintessence, Copenhagen, pp. 255–268.
149. Behera L, Kar I and Elitzur C (2006). In: Tuszynski JA (ed), The Emerging Physics of Consciousness. Springer, Berlin, pp. 327–350.
150. Resnick MD, Bearman PS, Blum RW et al (1997). Protecting adolescents from harm. Findings from the National Longitudinal Study on Adolescent Health. JAMA 278: 823–832.
151. Smirniotopoulos J (2000). Radiological Society of North America. Categorical Course in Diagnostic Radiology Neuroradiology Syllabus, Bethesda, Meryland.
152. Spence C (2002). Behav. Brain Res. 135(1–2):57–64.
153. Cabeza R and Kingstone A (2001). Handbook of Functional Neuroimaging of Cognition. MIT Press, Cambridge, MA.
154. Gibson JJ (1967). In: Bosma JF (ed), Symposium on Oral Sensation and Perception. Charles C. Thomas, Springfield, IL, pp. 111–136.
155. Johansson RS, Trulsson M, Olson KA, Westberg KG (1988). Exp. Brain Res. 72:204–208.
156. Trulsson M and Essick GK (1997). J. Neurophysiol. 77:737–748.
157. Rath EM and Essick GK (1990). Oral Maxillofac. Surg. 48:1181–1190
158. Petrovic AG and Stutzmann J (1977). In: McNamara J Jr (ed), The Biology of Occlusal Development. Monograph No. 7, Craniofacial Growth Series. University of Michigan, pp. 255–291.
159. Johansson RS and Valdo AB (1983). Trends Neurosci. 6:27–32.
160. Thyrow RC (1966). Edgewise Orthodontics, 2nd edition. C.V. Mosby, St. Louis, pp. 99.
161. Mason RM (1979). In: Oral Motor Behavior. U.S.A. DHEW Publ. No (NIH) 79-1845, pp. 32–48.

162. Thilander B (1961). Trans. Roy. Sch. Dent. Stockholm Umea 7:1–67.
163. Storey A (1976). In: Anderson DJ and Matthews B (eds), Mastication. John Wright and Sons, Bristol, pp. 50–57.
164. Siegel B (1996). The World of the Autistic Child: Understanding and Treating Autistic Spectrum Diseases. Oxford University Press, New York.
165. Miles TS (2004). In: Miles TS, Nauntofte B and Svensson P (eds), Clinical Oral Physiology. Quintessence, Copenhagen, pp. 199–217.
166. Barnard RI, Edgerton VR and Peter JB (1970). J. Appl. Physiol. 28:762–766.
167. Pimenidis MZ and Gianelly AA (1977). Am. J. Orthod. 71:94–102.
168. Eriksson PO and Thornell LE (1983). Arch. Oral Biol. 28:781–795.
169. Kraus ST (1988). TMJ Disorders. Churchill Livingstone, New York.
170. Moller E, Sessle BJ and Hannam AG (1976). Mastication and Swallowing. Biological and Clinical Correlates. University of Toronto Press, Toronto, pp. 128–144.
171. Ahlgren J (1976). In: Anderson DJ and Matthews B (eds), Mastication. John Wright and Sons, Bristol, pp. 119–130.
172. Hannam AG (1979). U.S.A DHEW Publ. No (NIH) 79-1845, pp. 87–118.
173. Watt DM (1976). In: Anderson DJ and Matthews B (eds), Mastication. John Wright and Sons, Bristol, pp. 242–250.
174. Sato S (1994). Angle Orthod. 64 (2):105–112.
175. Moss ML and Salentijn L (1969). Am. J. Orthod. 55:556–577.
176. Moss ML and Cowin SC (1995). In: Lanza R, Langer R and Chick W (eds), Textbook of Tissue Engineering. Springer, New York.
177. Harvold EP (1974). The Activator in Interceptive Orthodontics. C.V. Mosby. St. Louis.
178. Lindegard B (1953). Acta Psychiat. Scand. Suppl. 86.
179. Björk A (1960). In: Lundström A (ed), Introduction to Orthodontics. Mc Graw-Hill, London, pp. 104–140.
180. Solow B(1966). Acta Odont. Scand. 24(suppl):46.
181. Sassoumi V (1969). Am. J. Orthodont. 55:109–123.
182. Moller E (1966). Acta Phyciol. Scand. 69(suppl):280.
183. Ahlgren J (1966). Acta Odont. Scand. 24(suppl):44.
184. Ingervall B and Thilander B (1975). Angle Orthodont. 45:249.
185. Rasmussen OC and Moller E (1975). J. Dent. Res. 54(special issue A):L 94.
186. Stelmach GE (1979). U.S.A DHEW Publ. No (NIH) 79–1845, pp. 59–76.
187. Weiss PA (1968). Dynamics of Development: Experiments and Inferences. Academic Press, New York.
188. Dellow PG (1976). In: Sessle BJ and Hannam AG (eds), Mastication and Swallowing. Biological and Clinical Correlates. University of Toronto Press, Toronto, pp. 6–21.
189. Taylor A, Stephens JA, Somjen G and Harrison LM (1978). In: Perryman J (ed), Oral Physiology and Occlusion. An International Symposium. Pergamon Press, New York.
190. Hughes GM (1975). Comparative Physiology of Vertebrate Respiration. Harvard University Press, Cambridge, MA.
191. Sessle BJ (1978). In: Perryman JH (ed), Oral Physiology and Occlusion. An International Symposium. Pergamon Press, New York, pp. 91–114.

192. Goldberg LJ (1978). In: Perryman JH (ed), Oral Physiology and Occlusion. An International Symposium. Pergamon Press, New York, pp. 78–90.
193. Miller AJ and Vargervik K (1979). In: McNamara JA Jr and Ribbens KA (eds), Naso-Respiratory Function and Craniofacial Growth. Monograph Number 9. Craniofacial Growth Series. Center for Human Growth and Development. University of Michigan, Ann Arbor, Michigan.
194. King C (2006). In: Tuszynski JA (ed), The Emerging Physics of Consciousness. Springer, Berlin, pp. 407–456.
195. Humphrey T (1970). In: Bosma JF (ed), Symposium on Oral Sensation and Perception. Charles C. Thomas, Springfield, IL, pp. 195–233.
196. Moss ML (1972). Am. J. Orthod. 61:479–485.
197. Nakamura Y, Kubo Y, Nozaki S, Enomoto S and Katoh M (1981). In: Kawamura Y and Dubner R (eds), Oral-Facial Sensory and Motor Functions. Quintessence, Tokyo, pp. 37–44.
198. Proffit WR (1978). In: Perryman JH (ed), Oral Physiology and Occlusion. An International Symposium. Pergamon Press, New York, pp. 55–77.
199. Amen DG (1998). Change Your Brain Change Your Life. Three Rivers Press, New York.
200. Bardow A, Pedersen AML and Nauntofte B (2004). In: Miles TS, Nauntofte B and Svensson P (eds), Clinical Oral Physiology. Quintessence, Copenhagen, pp. 17–51.
201. Kallman F and Grobstein C (1965). Develop. Biol. 11:169.
202. Copenhaver WM (1971). In: Willier B, Weiss PA and Hamburger V (eds), Analysis of Development. Hafner, New York, pp. 440–461.
203. Weiss PA (1971). In: Willier B, Weiss PA and Hamburger V (eds), Analysis of Development. Hafner, New York, pp. 346–401.
204. Hendry IA (1976). In: Ehrenpreis S and Kopin IJ (eds), Reviews of Neuroscience. Raven Press, New York, pp. 149–194.
205. Levi-Montalcini R (1997). The saga of the Nerve Growth Factor. World Scientific, Singapore.
206. Linder-Aronson S and Woodside DG (2000). Excess Face Height Malocclusion: Etiology, Diagnosis, and Treatment. Quintessence, Chicago.
207. Aslanidi OV and Mornev OA (1966). International Conference on Mathematical Models of Nonlinear Excitation Processes. Tver, Russia, July 2–5.
208. Davydov AS (1982). Biology and Quantum Mechanics. Pergamon Press, Oxford.
209. Duke T and Jülicker F (2003). Phys. Rev. Lett. 90:158101.
210. Rhode WS and Recio A (2001). J. Acoust. Soc. Am. 110:3140–3154.
211. Moore BCJ (2003). An Introduction to the Psychology of Hearing. Academic Press, Boston.
212. Eccles JC (1966). In: Granit R (ed), Muscular Afferents and Motor Control. First Nobel Symposium. Wiley, New York, pp. 19–36.
213. Glaser GH and Higgins DC (1966). In: Granit R (ed), Muscular Afferents and Motor Control. First Nobel Symposium. Wiley, New York, pp. 121–138.
214. Bowman JP Jr (1971). The Muscle Spindle and Neural Control of the Tongue. Charles C. Thomas, Springfield, IL.
215. Matthews PBC (1964). Physiol. Rev. 44:219.

216. Oscarsson O (1966). In: Granit R (ed), Muscular Afferents and Motor Control. First Nobel Symposium. Wiley, New York, pp. 307–316.
217. Türker KS (2002). Crit. Rev. Oral Biol. Med. 13:85–104.
218. Bakke M (1993). Scand. J. Dent. Res. 101:314–431.
219. Lund JP, Kolta A, Westberg KG and Scott G (1998). Curr. Opin. Neurobiol. 8:718–724.

Index

MIX
Papier aus verantwortungsvollen Quellen
Paper from responsible sources
FSC® C105338

If you have any concerns about our products,
you can contact us on
ProductSafety@springernature.com

In case Publisher is established outside the EU,
the EU authorized representative is:
Springer Nature Customer Service Center GmbH
Europaplatz 3, 69115 Heidelberg, Germany

Printed by Libri Plureos GmbH
in Hamburg, Germany